ELECTRICAL AND ENGINEERING DEVELOPMENTS

ELECTRIC POWER SYSTEMS IN TRANSITION

ELECTRICAL AND ENGINEERING DEVELOPMENTS

ELECTRICAL AND ENGINEERING DEVELOPMENTS

ELECTRIC POWER SYSTEMS IN TRANSITION

OLIVIA E. ROBINSON
EDITOR

Nova
Nova Science Publishers, Inc.
New York

NOTICE TO THE READER

The Publisher has taken reasonable care in the preparation of this book, but makes no expressed or implied warranty of any kind and assumes no responsibility for any errors or omissions. No liability is assumed for incidental or consequential damages in connection with or arising out of information contained in this book. The Publisher shall not be liable for any special, consequential, or exemplary damages resulting, in whole or in part, from the readers' use of, or reliance upon, this material. Any parts of this book based on government reports are so indicated and copyright is claimed for those parts to the extent applicable to compilations of such works.

Independent verification should be sought for any data, advice or recommendations contained in this book. In addition, no responsibility is assumed by the publisher for any injury and/or damage to persons or property arising from any methods, products, instructions, ideas or otherwise contained in this publication.

This publication is designed to provide accurate and authoritative information with regard to the subject matter covered herein. It is sold with the clear understanding that the Publisher is not engaged in rendering legal or any other professional services. If legal or any other expert assistance is required, the services of a competent person should be sought. FROM A DECLARATION OF PARTICIPANTS JOINTLY ADOPTED BY A COMMITTEE OF THE AMERICAN BAR ASSOCIATION AND A COMMITTEE OF PUBLISHERS.

LIBRARY OF CONGRESS CATALOGING-IN-PUBLICATION DATA
Electric power systems in transition / editors, Olivia E. Robinson.
p. cm.
  Includes bibliographical references and index.
  ISBN 978-1-61668-985-8 (hardcover)
 1. Electric power systems. I. Robinson, Olivia E.
  TK1001.E425 2010
  621.319'1--dc22
2010014123

Published by Nova Science Publishers, Inc. ✤ *New York*

CONTENTS

PREFACE

The electric power system industry is undergoing massive changes around the world. Despite the changes with different structures, market rules, and uncertainties, an energy management system (EMS) control center must always be in place to maintain the security, reliability, and quality of electric service. The function of state estimation is reviewed in this book and is becoming more important, because it is the primary tool for monitoring and control based on real-time data received from measurement units. Also discussed in this book, are the design and application criteria for a proposed overcurrent relay in radial distribution networks; load modeling in power systems; wide-area stability and voltage control of large interconnected power systems and others.

Chapter 1- Frequency-dependent transmission line models can be developed in phase domain or in mode domain. Associated to these transmission line models, time domain or frequency domain can be used for electromagnetic transient analysis solutions [1-6]. Because of these options, there are several models for transmission line representation in electromagnetic transient studies [7-12]. One reason for this is the longitudinal parameter frequency dependence [3-6, 11-19]. With the aim to improve the frequency dependence representation, several models have been suggested [1-19]. In some cases, phase-mode transformation is applied, considering the problem in mode domain, and searching for improving the frequency-dependent parameter representation [3-6, 8-9]. On the other hand, changes in the system structure as well as voltage and current surge waves are better modeled on the time domain. So, a hybrid model based on phase-mode transformation can be applied using a better line parameter frequency-dependence representation with the time domain model advantages.

In exact mathematical development, the phase mode transformation matrices depend on the frequency because the line parameters are frequency dependent. Because of this, all values that represent the electrical line characteristics, such as the longitudinal impedance (Z) and transversal admittance (Y) matrices, are influenced by frequency [3, 11-19]. Using the exact phase-mode transformation, we calculated the line parameters. It is an alternative methodology to calculate transmission line parameters per unit length. With this methodology the transmission line parameters can be obtained from impedances measured in one terminal of the line. So, a new procedure is shown to calculate frequency-dependent transmission line parameters directly from currents and voltages from the line terminals.

Chapter 2- Modelling the power systems load is a challenge since the load level and composition varies with time. An accurate load model is important because there is a substantial component of load dynamics in the frequency range relevant to system stability. The composition of loads need to be characterized because the time constants of composite loads affect the damping contributions of the loads to power system oscillations, and their effects vary with the time of the day, depending on the mix of motors loads.

This chapter has two main objectives: 1) describe the load modeling in small signal using on-line measurements; and 2) present a new approach to develop models that reflect the load response to large disturbances.

Small signal load characterization based on on-line measurements allows predicting the composition of load with improved accuracy compared with post-mortem or classical load models. Rather than a generic dynamic model for small signal modelling of the load, an explicit induction motor model is used so the performance for larger disturbances can be more reliably inferred. The relation between power and frequency/voltage can be explicitly formulated and the contribution of induction motors extracted. One of the main features of this work is the induction motor component can be associated to nominal powers or equivalent motors.

When large disturbances are considered, loads play a major role in the recovery of the voltage. Among the different types of loads, induction motors have a extremely non-linear characteristic. After a disturbance, an induction motor can be reaccelerated, stalled or tripped. The stalled case is the most important case because of the large amount of reactive power demanded, which can influence the recovery times of voltage after the fault is cleared and has the potential to drive the system to voltage collapse. Regarding the importance of induction motors and the characterization of the pre-fault state using on-line measurements, a new model for large disturbance is proposed in this work. It will be shown that information contained in the on-line measurements allows description of the static and induction motor components in the pre-fault state. The induction motor component can be described as groups of motors with similar inertia and nominal power, which can be obtained from the small signal model. To obtain a complete characterization of the response of the load to large disturbances, two indices are proposed in this work: the severity index and the tripping index. The first one will describe the probability of motors in a group stall after the fault and the second describes the amount of load that will be lost in a major disturbance. Both indexes are directly associated with the recovery of the voltage. The importance of the indices is that they can be identified through a limited set of fault measurements while a full response model across the full range of depth and duration of disturbances would take hundreds of fault records.

Chapter 3- Wide-area stability and voltage control of large interconnected power systems has attracted considerable interest in the last two decades. Advances in signal processing algorithms, along with continuously growing computational resources and the evolution in protection schemes, are beginning to make feasible the analysis and control of intersystem oscillation problems using measurement-based, wide-area protection systems.

A considerable amount of research has been carried out over the last twenty years to develop advanced, wide-area monitoring, protection and control systems. Conventional stability ad voltage control systems may be inadequate to cope with wide-area stability problems and emergency conditions.

Integrated synchro-phasor measurements are emerging as the backbone for future real-time wide-area monitoring, protection and control systems. The advent of remotely sensed data from satellites greatly extends the scales of time and space at which control can be carried out. Efficient utilization of these technologies, however, poses enormous practical and theoretical challenges that have to be addressed especially under deregulated power system operation.

This chapter discusses the experience in the design and evaluation of advanced wide-area stability and voltage control systems to remove transmission constraints in the system. Examples taken from a large power system show that wide-area Special Protection Systems (SPSs) involving direct tripping of generation (load) and wide-area measurement-based secondary voltage control can maintain system stability and prevent from further system wide spread cascading outages. Control options using both, shunt and series compensation schemes as well as under voltage load shedding are evaluated

Implementation details using WAMS/SCADA systems are also discussed.

Chapter 4- This chapter contains the design and application criteria for a proposed overcurrent relay. This relay uses independent functions to detect faults and to calculate the operation time. Two main functions are proposed, the first function offers a rather faster operation for low faults currents and greater sensitivity and the second function provides sensitivity for the detection of high impedance faults (HIF). The proposed relay have a negative sequence detector and positive sequence detector for confirm the presence of a fault, thus the negative sequence relay coordination is not necessary. The line protection scheme is simplified. The functional changes introduced simplify the setting process by the user with a minimal change in the relay's firmware and without change of relay hardware.

Chapter 5- The electric power industry is undergoing massive changes around the world. Despite the changes with different structures, market rules, and uncertainties, an energy management system (EMS) control center must always be in place to maintain the security, reliability, and quality of electric service [1-2]. It means that EMS in the open energy market must respond quickly, reliably and efficiently to the market changes.

In this case, the function of state estimation (SE) is becoming more important, because it is the primary tool for monitoring and control based on the real-time data received from the measurement units. State estimation takes all the telemetry seen so far and uses it to determine the underlying behavior of the system at any point in time. It contributes to excluding errors in meters and compensating meter deficiency due to some faults with other measurements. As a result, it is possible for system operators to understand the state of the system through state estimation appropriately. This chapter introduces the methods of state estimation including SE with phasor measurement unit (PMU) and harmonic state estimation.

Chapter 6- This article presents a general approach for locating any type of short circuit faults on a double-circuit transmission line. By making use of the bus impedance matrix technique, voltage measurements at one or two buses are utilized as inputs, which may be distant from the faulted section. The bus impedance matrix of each sequence network with addition of a fictitious bus at the fault point can be constructed as a function of fault location by drawing on network analysis. Fault location can then be obtained based on bus impedance matrix and boundary conditions of different fault types. It is assumed that the network data are available. Quite accurate results have been achieved based on simulation studies.

In: Electric Power Systems in Transition
Editor: Olivia E. Robinson, pp. 1-73

ISBN: 978-1-61668-985-8
© 2010 Nova Science Publishers, Inc.

Chapter 1

PHASE-MODE TRANSFORMATION MATRIX APPLICATION FOR TRANSMISSION LINE AND ELECTROMAGNETIC TRANSIENT ANALYSES

*Afonso José do Prado[1], Sérgio Kurokawa[1],
José Pissolato Filho[2], Luiz Fernando Bovolato[1]
and Eduardo Coelho Marques da Costa[2]*

[1]Electrical Engineering Department – DEE, The University of São Paulo State – UNESP,
Brazil
[2]Electrical Engineering Department – DSCE, The State University of Campinas –
UNICAMP, Brazil

INTRODUCTION

Frequency-dependent transmission line models can be developed in phase domain or in mode domain. Associated to these transmission line models, time domain or frequency domain can be used for electromagnetic transient analysis solutions [1-6]. Because of these options, there are several models for transmission line representation in electromagnetic transient studies [7-12]. One reason for this is the longitudinal parameter frequency dependence [3-6, 11-19]. With the aim to improve the frequency dependence representation, several models have been suggested [1-19]. In some cases, phase-mode transformation is applied, considering the problem in mode domain, and searching for improving the frequency-dependent parameter representation [3-6, 8-9]. On the other hand, changes in the system structure as well as voltage and current surge waves are better modeled on the time domain. So, a hybrid model based on phase-mode transformation can be applied using a better line parameter frequency-dependence representation with the time domain model advantages.

In exact mathematical development, the phase mode transformation matrices depend on the frequency because the line parameters are frequency dependent. Because of this, all values that represent the electrical line characteristics, such as the longitudinal impedance (Z) and transversal admittance (Y) matrices, are influenced by frequency [3, 11-19]. Using the exact phase-mode transformation, we calculated the line parameters. It is an alternative methodology to calculate transmission line parameters per unit length. With this methodology the transmission line parameters can be obtained from impedances measured in one terminal of the line. So, a new procedure is shown to calculate frequency-dependent transmission line parameters directly from currents and voltages from the line terminals.

If the line is a symmetric three-phase line, the phase-mode transformation can be carried out using a 3-order real and constant matrix and a 2-order frequency dependent transformation matrix. For analyses and simulations, the computational time can be decreased because the order of frequency dependent matrix is reduced. Also using the exact phase-mode transformation matrices and considering the frequency range mentioned above, an alternative procedure is carried out for the equivalent conductor determination of a bundle of subconductors. Considering symmetrical bundles, the proposed alternative obtains results that are similar to those obtained from the geometric mean radius (GMR) procedure. For asymmetrical bundles, the alternative procedure based on phase-mode transformations is more accurate when frequency values in the range from 10 Hz to 100 Hz are considered.

If the mathematical model based on the exact phase-mode transformation is applied to digital programs, the result can be a slow digital routine for transmission line analyses. An alternative that can be considered is the use of a single real transformation matrix. It is a way to obtain fast transmission line transient simulations as well as to avoid convolution procedures in this simulation type [4-6, 8-10]. Clarke's matrix has presented interesting performances for three-phase transmission lines: exact results for transposed cases and negligible errors for non-transposed ones [7]. This matrix is single, real, frequency independent, and identical for voltage and current. So, it is analyzed the changing the exact phase-mode transformation matrices into matrices composed of constant and real elements. The phase-mode transformation matrix applications can lead to complicated and slow numeric routines. An alternative is the single real matrix use for a wide frequency range. It is possible because the elements of the eigenvectors that are the phase-mode transformation matrices smoothly vary in function of frequency. In this case, Clarke's matrix is used and checked for symmetrical and asymmetrical three-phase transmission lines considering transposed and untransposed cases. For transposed cases, Clarke's matrix is an eigenvector of the three-phase transmission lines. For untransposed cases, this matrix is a good approximation to the exact transformation considering the eigenvalue comparisons. However, this is not true for off-diagonal elements of the matrix obtained from the application of Clarke's matrix. Using a correction procedure, the elements of Clarke's matrix are corrected leading to two new transformation matrices: one related to the voltage values and the other related to the current values. Because the obtained errors are negligible and the mentioned off-diagonal elements become negligible values, these new matrices can be considered eigenvectors of the analyzed lines. These matrices smoothly vary in function of frequency, and their elements have small imaginary part. They are checked using untransposed symmetrical and untransposed asymmetrical three-phase transmission lines.

Finally, for future development, the numeric routines obtained in both parts of this chapter will be applied with state variables for electromagnetic transient simulations. So, it is suggested a model for transmission line that considers the frequency influence using state variables. Another suggestion for future development is searching adequate transformation matrices for systems with some parallel three-phase circuits.

I. TRANSMISSION LINE PARAMETERS DERIVATION FROM IMPEDANCE VALUES [20]

The self and mutual impedances present in the overhead transmission line equations in the frequency domain can be derived from the solution of Maxwell's equations for the boundary conditions at the contact surfaces of the three relevant materials: conductor, air and ground. [21]. Evaluating line parameters, there are some usual assumptions that imply in physical approximations related to line geometry or electromagnetic field behavior [22]. The main line geometric approximations are: the soil surface is a plane; the line cables are horizontal and parallel among themselves; the distance between any pair of conductors is much higher than the sum of their radii; the electromagnetic effects of structures and insulators are neglected. About electromagnetic field behavior for line transversal parameters, it is assumed the quasi-stationary electromagnetic field simplification [22]. The most used procedures that represent the ground assume a constant and frequency independent ground conductivity, neglecting the ground dielectric permittivity. There are situations where the mentioned simplifications can not be assumed. An example of a situation in which physical properties can be changed is in the study of electromagnetic transients that include nonuniform lines that could be portions of transmission lines where the conductors are not parallel [23].

The transmission line parameters could be derived from measured transmission line impedances, but there are some practical difficulties to measure the frequency response of a transmission line [24]. It requires a strong motivation to get the consent from a power utility to make an outage of a long transmission line. The experimental setup may be nontrivial. Frequency measurement requires a voltage source with variable frequency and high power. Due to difficulties above mentioned the procedure developed in this item has been used in transmission lines represented by digital models. So, it proposes a methodology to calculate longitudinal and shunt parameters per unit length of overhead transmission lines from the impedances measured at line terminal.

I.1. Frequency-Dependent Transmission Line Parameter Calculating

One of the most important aspects for modeling of electromagnetic transient in transmission lines is that the line parameters depend on the frequency. Models that assume constant parameters have not adequately simulated the response of the line over the wide range of frequency for transient conditions. The constant line parameter representation produces a magnification of the higher harmonics, a general distortion of the wave shapes and exaggerated magnitude peaks [17].

Using the frequency domain, the self and mutual impedances included in the overhead transmission line equations can be derived from the solution of Maxwell's equations. In these equations, the impedance matrix (Z) determination falls into three parts: internal longitudinal, external longitudinal and soil effect impedances; or two parts: self and mutual impedances. The internal longitudinal impedance is associated with the electromagnetic field within the conductor. Electromagnetic fields do not affect mutual terms and the mutual impedance does not depend on the internal longitudinal impedance. Due to the skin effect, the resistance increases whereas the inductance decreases. The external longitudinal impedance is associated with the electromagnetic field outside the conductors. In this case, it is assumed a lossless ground and the other assumptions indicated before. For a lossy ground, it is considered the soil effect that means an additional parcel of the external longitudinal impedance matrix. A similar analysis applies also to transversal admittance matrix (Y) where, for typical conditions, it is reasonable to assume ideal conductors and ground up to 1 MHz.

The parameters of transmission lines with ground return are highly dependent on the frequency. Formulas to calculate the influence of the ground return were developed by Carson and Pollaczek and these formulas can also be used for power lines. Both lead to identical results for overhead lines, but Pollaczek's formula is more general inasmuch as it can also be used for underground conductors or pipes [1, 2].

I.2. Line Parameter Calculating from Phase Current and Voltage Values

The basic equations of a transmission line for sinusoidal alternated electrical magnitude complex representation are [22]:

$$\frac{d^2 V_{PH}}{dx^2} = Z \cdot Y \cdot V_{PH} \quad and \quad \frac{d^2 I_{PH}}{dx^2} = Y \cdot Z \cdot I_{PH} \tag{1}$$

The Z and Y matrices are per unit length longitudinal impedance and shunt admittance matrices, respectively. The elements of these matrices are frequency dependent. The V_{PH} and I_{PH} vectors are, respectively, transversal line voltage and longitudinal line current vectors. These equations are valid if the electromagnetic field has a quasi-stationary behavior in orthogonal direction to line axis [22].

Poly-phase transmission line equations can be solved transforming n coupled equations into n decoupled equations. Decoupling of equations is carried out using a suitable chosen modal transformation matrix T_I changing the YZ matricial product into its diagonal form [5, 11]:

$$T_I^{-1} \cdot Y \cdot Z \cdot T_I = \lambda \tag{2}$$

The λ matrix is the eigenvalue one.

Substituting equation (2) in equation (1), it is obtained the basic equations of a transmission line in mode domain [14]:

$$\frac{d^2 V_M}{dx^2} = Z_M \cdot Y_M \cdot V_M \quad and \quad \frac{d^2 I_M}{dx^2} = Y_M \cdot Z_M \cdot I_M \tag{3}$$

The Z_M and Y_M matrices are described as following [11]:

$$Z_M = T_I^T \cdot Z \cdot T_I \quad and \quad Y_M = T_I^{-1} \cdot Y \cdot T_I^{-T} \qquad (4)$$

The T index identifies the transposition of the analyzed matrix. The negative T index is related to the inverse transposed matrix. The T_I^{-1} is the inverse T matrix. The V_M and I_M vectors are, respectively, transversal line voltages and longitudinal line currents in mode domain.

Because matrices Z_M and Y_M are diagonal matrices, the $Z_M Y_M$ and $Y_M Z_M$ matricial products are diagonal matrices and there are no couplings among modes. For a generic mode, it is carried out [14]:

$$\begin{cases} E_A = E_B \cdot \cosh(\gamma \cdot d) - I_B \cdot Z_C \cdot \sinh(\gamma \cdot d) \\ I_A = -I_B \cdot \cosh(\gamma \cdot d) + \dfrac{E_B}{Z_C} \cdot \sinh(\gamma \cdot d) \end{cases} \qquad (5)$$

E_A and E_B are, respectively, modal voltages at terminals A and B in Figure I.1. The terms I_A and I_B are modal currents at these terminals and d is the line length. The terms γ and Z_C are, respectively, the propagation function and the characteristic impedance of the analyzed mode.

The Z_C and γ values are written as being [21]:

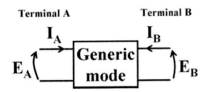

Figure I.1. Quadripole of a transmission line generic mode

$$\gamma = \sqrt{Z_{KK} \cdot Y_{KK}}$$

$$Z_C = \sqrt{\dfrac{Z_{KK}}{Y_{KK}}} \qquad (6)$$

In this case, Z_{KK} is an element of the Z matrix that represents the impedance of the k–mode. Y_{KK} is an element of the Y matrix that represents the impedance of the k-mode. The k–mode is the generic mode mentioned in Figure I.1.

Supposed a Z_L impedance load in terminal B shown in Figure I.1. The Z_L value is known and, using equation (5), it is possible to determine the Z_{EC} impedance of the generic mode equivalent circuit [23]. The Z_{EC} is described as:

$$Z_{EC} = \frac{Z_L \cdot \cosh(\gamma \cdot d) + Z_C \cdot \sinh(\gamma \cdot d)}{\cosh(\gamma \cdot d) + \dfrac{Z_L}{Z_C} \cdot \sinh(\gamma \cdot d)} \qquad (7)$$

Considering two specific line configurations, there are two impedances for the mode shown in Figure I.1. The first impedance is defined considering that terminal B is opened and the other impedance is defined considering terminal B is short-circuited. Z_{OPEN} is the equivalent impedance when terminal B is open ($Z_L \rightarrow \infty$), E_{AOPEN} is the voltage in terminal A when terminal B is open. I_{AOPEN} is the current related to this described line configuration. Manipulating equation (7), it is possible to write each Z_{OPEN} mode as function of γ and Z_C as following:

$$Z_L \rightarrow \infty \Rightarrow Z_{EC} = Z_{OPEN} = \frac{E_{AOPEN}}{I_{AOPEN}} = Z_C \cdot \tanh(\gamma \cdot d) \qquad (8)$$

Using a similar manipulating, it is obtained:

$$Z_L = 0 \Rightarrow Z_{EC} = Z_{CC} = \frac{E_{ACC}}{I_{ACC}} = Z_C \cdot \coth(\gamma \cdot d) \qquad (9)$$

In this case, Z_{CC} is the equivalent impedance when terminal B is short-circuited. For short-circuit, Z_L is null. E_{ACC} and I_{ACC} are, respectively, the voltage and current values in terminal A for this line configuration.

The Z_{OPEN} and Z_{CC} equivalent impedances can be calculated directly from currents and voltages of the line or directly from γ and Z_C. Using γ and Z_C related to Z_{CC} and Z_{OPEN} values, it is possible to calculate longitudinal and transversal transmission line parameters from equation (6).

It is considered a polyphase transmission line (n phases) where the sending ending terminal is called terminal A and the receipting ending terminal is called terminal B.. It is also considered that it is possible to obtain in these line terminals the V_{PH}, $I_{PH\ OPEN}$ and $I_{PH\ CC}$ vectors. The V_{PH} vector is composed by the sources connected in terminal A of the line. The $I_{PH\ OPEN}$ and $I_{PH\ CC}$ vectors have the longitudinal currents in terminal A of each phase of the line, considering terminal B opened and short-circuited, respectively. The mentioned vectors are written in modal domain as being:

$$\begin{aligned} E_M &= T_I^T \cdot V_{PH} \\ I_{M\ OPEN} &= T_I^{-1} \cdot I_{PH\ OPEN} \\ I_{M\ CC} &= T_I^{-1} \cdot I_{PH\ CC} \end{aligned} \qquad (10)$$

E_M is the vector with modal voltage sources. Each modal voltage is connected to terminal A of the respective mode of the line. The $I_{M\ OPEN}$ and $I_{M\ CC}$ vectors are current vectors in terminal A of each mode considering that terminal B is open and short-circuited, respectively. Considering n modes of a polyphase line, the equations (8) and (9) are used for calculating the equivalent impedances Z_{OPEN} and Z_{CC} of each mode. After that, manipulating this equation, it is obtained:

$$\coth(\gamma \cdot d) = \sqrt{\frac{Z_{OPEN}}{Z_{CC}}} \qquad (11)$$

The Z_{OPEN} and Z_{CC} equivalent impedances are known and obtained from equations (8) and (9), respectively. Manipulation the last equation, it is obtained the propagation function (γ) of a generic mode of the line.

On the other hand, the last equation can be write as following:

$$\coth(\gamma \cdot d) = \frac{e^{\gamma \cdot d} + e^{-\gamma \cdot d}}{e^{\gamma \cdot d} - e^{-\gamma \cdot d}} \qquad (12)$$

Equaling the equations (11) and (12), it is obtained:

$$\frac{e^{\gamma \cdot d} + e^{-\gamma \cdot d}}{e^{\gamma \cdot d} - e^{-\gamma \cdot d}} = \sqrt{\frac{Z_{OPEN}}{Z_{CC}}} \qquad (13)$$

From equation (13), it is possible to express γ as:

$$\gamma = \frac{\ln(F_1) + j \cdot \arccos(F_2)}{2d} \qquad (14)$$

F_1 and F_2 are determined by:

$$F_1 = \frac{(1 + D_1)^2 + D_2^2}{\sqrt{(1 - D_1^2 - D_2^2)^2 + 4D_2^2}} \qquad (15)$$

$$F_2 = \frac{D_1^2 + D_2^2 - 1}{\sqrt{(1 - D_1^2 - D_2^2)^2 + 4D_2^2}}$$

The D_1 and D_2 values are related to Z_{OPEN} and Z_{CC} as following:

$$\sqrt{\frac{Z_{OPEN}}{Z_{CC}}} = D_1 + j \cdot D_2 \qquad (16)$$

For determination of D_1 and D_2 values, equations (14) and (16) can be used in a specific frequency value applying the following restriction:

$$D_1 \neq \pm 1 \quad and \quad D_2 \neq 0 \qquad (17)$$

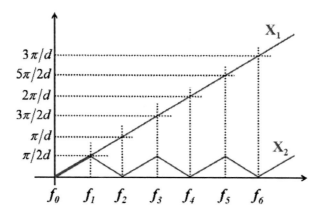

Figure I.2. The γ function imaginary part for a generic mode

The last restriction is necessary for mode of the analyzed line in each specific frequency used for the γ calculating.

However, it can be observed that there is a no negligible difference between the imaginary part of the equation (14) and the correspondent value obtained from the equation (6). It is analyzed from the next equations, considering a generic mode.

$$X_1 = \text{Im}\left(\sqrt{Z_{KK} \cdot Y_{KK}}\right)$$

$$X_2 = \frac{\arccos(F_2)}{2d}$$

(18)

The graphic shown in Figure I.2 compares the X_1 and X_2 values that depend on the frequency.

Functions X_1 and X_2 are equal for frequencies between f_0 and f_1 only. For the other frequency ranges in the used graphic, there are increasing differences between the both functions. However, it is possible to equal X_1 and X_2 for other frequency ranges using a single procedure. Consider the H function as shown in the next equation.

$$H(f_K) = \left|\frac{\partial X_2(f_K)}{\partial f}\right|(f_K - f_{K-1}) + H(f_{K-1})$$

(19)

In this case, it can be used the following approximation:

$$\frac{\partial X_2(f_K)}{\partial f} \approx \frac{X_2(f_K) - X_2(f_{K-1})}{(f_K - f_{K-1})}$$

(20)

Equations (19) and (20) can be used if f_K is higher than f_1. If f_K is lower than f_1, H(f) is written as following:

$$H(f_K) = X_2(f_K)$$

(21)

Using equations (19), (20) and (21), it is obtained a function relating X_1 and X_2 for each frequency.

Substituing the equation (6) in product of (8) and (9), it is obtained the following:

$$\frac{Z_{KK}}{Y_{KK}} = Z_{CC} \cdot Z_{OPEN} \qquad (22)$$

Using equations (6) and (22), it is determined:

$$Z_{KK} = \gamma \sqrt{Z_{CC} \cdot Z_{OPEN}} \qquad (23)$$

$$Y_{KK} = \frac{\gamma}{\sqrt{Z_{CC} \cdot Z_{OPEN}}}$$

The terms Z_{OPEN} and Z_{CC} are calculated from currents and voltages obtained in one terminal of a generic mode considering that the other terminal is open and short-circuited, respectively. Applying the proposed development, the modal propagation function of each mode is calculated, considering a frequency range of interest. From the modal propagation function values, longitudinal impedance and transversal admittance of each mode are derived. With longitudinal impedance and transversal admittance of each mode, longitudinal impedance matrix Z_M and transversal admittance matrix Y_M in mode domain are determined. These matrices are converted to the phase domain as following:

$$Z = T_I^{-T} \cdot Z_M \cdot T_I^{-1} \qquad (24)$$

$$Y = T_I \cdot Y_M \cdot T_I^{-T}$$

The Z and Y matrices have been defined in equation (1) as the longitudinal impedance and the shunt admittance ones in phase domain, respectively. So, if currents and voltages for the opened and short-circuited line are known, the Z and Y matrices can be calculated.

The procedure proposed in this section is based in the knowledge of the modal transformation matrix and the voltage and current magnitudes at the beginning of the line during open and short circuit conditions. These conditions can reduce the applicability of the methodology, if it is necessary to consider the exact values. If it is applied some approximations, the proposed procedure can be used obtaining adequate results for practical applications. The main restriction is that the modal transformation matrix should be known. For two-phase transmission line that actual samples are the transmission lines in continuous current, the modal transformation matrices are known for any frequency because these matrices are not influenced by the line geometric characteristics and the frequency.

For transposed three-phase transmission lines, considering symmetrical and asymmetrical cases, Clarke's matrix is an eigenvector matrix for any frequency value and it is independent of the line geometric characteristics. Considering untransposed three-phase transmission lines, there are some suggestions that can be applied. These are analyzed in the next items of this chapter.

II. AN ALTERNATIVE MODEL FOR EQUIVALENT CONDUCTOR DETERMINATION FROM BUNDLED CONDUCTORS

Connecting two or more sub-conductors, bundled conductors can be an efficient alternative to increase the capacity of high voltage transmission lines without conductor gauge increasing and maintaining the electromagnetic interference in acceptable levels. The bundled conductors are composed by sub-conductors connected in parallel and it is used spacers for attaching the bundle to the towers or along the sags among the line towers [25]. In actual systems, up to 230 kV, for the most of transmission lines, each phase is composed by a bundled conductor [26-28]. Using the mentioned spacers, the spacing among the adjacent sub-conductors is from 0.4 to 0.6 m for conventional lines. In these cases, the sub-conductors are equal and it can be considered that the current is equally distributed among the sub-conductors. These are the ideal conditions for the classical bundle conductor modeling application where the geometric mean radius (GMR) concept is applied. Using this concept, the bundle conductor is changed into an equivalent conductor at the geometric center of the bundle [29-35].

There are systems with non-symmetrical bundle conductors. There are also cases where the bundle conductor is composed by different sub-conductors. For these non conventional bundle conductor cases, the GMR concept is not appropriated and, because of this, an alternative procedure is introduced in this item, defining an equivalent conductor of a bundle conductor using the unbalanced distribution of the current among the sub-conductors.

II.1. Analyses for Single Conductors

It is considered a simple system with to single conductors in Figure II.1. Based on this system, the analyses can be similarly carried out increasing the number of conductors.

The radius of conductor i is r_i, as well as, r_k is the radius of conductor k. The system shown in Figure II.1 can be characterized using the longitudinal impedance (Z) and the shunt admittance (Y) matrices per unit length. The Z matrix is written as:

$$Z = R + j\omega L \qquad (25)$$

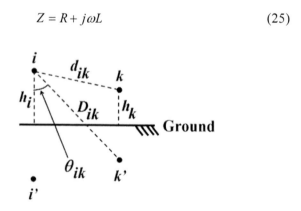

Figure II.1. System with two single conductors.

In the last equation, R is the longitudinal resistance matrix and L is the longitudinal inductance one. The R and L matrices are frequency dependent and also influenced by the skin and ground effects.

The Y real part matrix is the shunt conductance matrix (G) and the Y imaginary part matrix is the shunt capacitance matrix (C). For the most cases related to the power system conductors, the G matrix can be considered null and the C matrix can be considered frequency independent. The Y matrix is:

$$Y = G + j\omega C \cong j\omega C \tag{26}$$

Each self impedance (Z_{ii}), which is in the Z main diagonal, is written as adding three elements: the self external impedance ($Z_{ext\ ii}$), the self impedance due to the ground effect ($Z_{ground\ ii}$) and the self internal impedance ($Z_{int\ ii}$). Each mutual impedance (Z_{ik}), which is out of the Z matrix main diagonal, is written as adding two elements: the mutual external impedance ($Z_{ext\ ik}$) and the mutual impedance due to the ground effect ($Z_{ground\ ik}$). It is shown in the next equation.

$$Z_{ii} = Z_{ext\ ii} + Z_{ground\ ii} + Z_{int\ ii}$$
$$\tag{27}$$
$$Z_{ik} = Z_{ext\ ik} + Z_{ground\ ik}$$

The external impedances are calculated as demonstrated in the next equations written for the *i* conductor of Figure II.1. In this case, d_{ik} is the distance between the *i* and *k* conductors and D_{ik} is the distance between the *i* conductor and the *k* conductor image.

$$Z_{ext\ ii} = j\omega \frac{\mu_0}{2\pi} \ln\left(\frac{2h_i}{r_i}\right) \quad and \quad Z_{ext\ ik} = j\omega \frac{\mu_0}{2\pi} \ln\left(\frac{D_{ik}}{d_{ik}}\right) \tag{28}$$

The ground effects for impedances are calculated using the next equations that are also written basing on the *i* conductor. Carson's infinite series are applied for determination of ground effect impedances.

$$Z_{ground\ ii} = R_{ii}\left(a_{ii}, \varphi_{ii}\right) + j\omega L_{ii}\left(a_{ii}, \varphi_{ii}\right)$$
$$\tag{29}$$
$$Z_{ground\ ik} = R_{ik}\left(a_{ik}, \varphi_{ik}\right) + j\omega L_{ik}\left(a_{ik}, \varphi_{ik}\right)$$

Based on Figure II.1, it is determined the following values:

$$a_{ii} = 4\pi\sqrt{5}\cdot 10^{-4}\cdot h_i\sqrt{\frac{\omega}{2\pi\rho}} \qquad a_{ik} = 4\pi\sqrt{5}\cdot 10^{-4}\cdot D_{ik}\sqrt{\frac{\omega}{2\pi\rho}}$$
$$\tag{30}$$
$$\varphi_{ii} = 0 \qquad\qquad \varphi_{ik} = \theta_{ik}$$

Taking the i conductor as a sample, the self internal impedance is calculated basing on the radius of this conductor (r_i). The m value is determined by equation (32) depending on the permeability and resistivity of the analyzed conductor.

$$Z_{int\,ii} = \frac{j\omega\mu_i}{2\pi r_i}\left(\frac{ber(mr_i)+j\,bei(mr_i)}{ber'(mr_i)+j\,bei'(mr_i)}\right) \tag{31}$$

The m value in the last equation is calculated using the next equation where μ_i is the permeability of the conductor r_i and ρ_i is the resistivity of this conductor.

$$m = \sqrt{\frac{\omega\mu_i}{\rho_i}} \tag{32}$$

Neglecting the real part of the shunt admittance matrix, this matrix is equal to the shunt capacitance matrix that is calculated as:

$$C = P^{-1} \tag{33}$$

The P matrix is the Maxwell's potential coefficient one and every element of this matrix is calculating as following:

$$P_{ii} = \frac{1}{2\pi\varepsilon_0}\ln\left(\frac{2h_i}{r_i}\right) \quad and \quad P_{ik} = \frac{1}{2\pi\varepsilon_0}\ln\left(\frac{D_{ik}}{d_{ik}}\right) \tag{34}$$

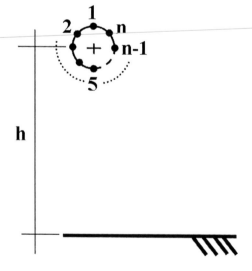

Figure II.2. A generic bundled conductor

II.2. Application of Geometric Mean Radius (GMR) Concept for Changing a Bundled Conductor into an Equivalent Conductor

A bundle conductor can be changed into an equivalent conductor. This equivalent conductor can be represented by the equivalent longitudinal impedance (Z_{EQ}) and the equivalent shunt admittance (Y_{EQ}). These values are determined using the same equations that are used for an individual conductor. So, the equivalent longitudinal impedance is composed by the external impedance, the impedance due to ground effect and the internal impedance. The shunt admittance is composed by the capacitance. It is shown in the next equation and based on Figure II.2 where h is the height of the equivalent conductor and corresponds to the height of the geometric center of the bundled conductor. It is considered n sub-conductors for this generic bundled conductor.

$$Z_{EQ} = Z_{EQ\,ext} + Z_{EQ\,ground} + Z_{EQ\,int}$$

(35)

$$Y_{EQ} = j\omega C_{EQ}$$

The equivalent external impedance is calculated as following:

$$Z_{EQ\,ext} = j\omega \frac{\mu_0}{2\pi} \ln\left(\frac{2h}{r_{GMR}}\right)$$

(36)

In this case, h is the height of the symmetrical center of the bundled conductor and the r_{GMR} is the geometric mean radius (GMR) of the bundle. This last value is defined as:

$$r_{GMR} = \sqrt[n \cdot n]{\prod_{i=1}^{n}\left(r_i \cdot \prod_{k=1}^{n} d_{ik}\right)}, \quad k \neq i$$

(37)

For the last equation, r_i is the i sub-conductor radius and d_{ik} is the distance between the sub-conductors i and k shown previously in Figure II.1.

Determining the equivalent ground impedance, the angle φ is null and the real and imaginary terms of the Z_{ground} value are based in the procedure used for individual conductors that is previously presented. It is represented as:

$$Z_{EQ\,ground} = R_{EQ}(a,\varphi) + j\omega L_{EQ}(a,\varphi)$$

(38)

Generally, the equivalent internal impedance is determined neglecting the shunt admittances of the sub-conductors. This value is obtaining using:

$$\frac{1}{Z_{EQ\,int}} = \sum_{i=1}^{n} \frac{1}{Z_{i-int}}$$

(39)

In case of equal sub-conductors, the last equation is simplified and it is based on the internal impedance of any sub-conductor of the bundle ($Z_{SUB\,int}$). It is obtained from the next equation:

$$Z_{EQ\,int} = \frac{Z_{SUB\,int}}{n} \tag{40}$$

The equivalent shunt admittance is equal to the shunt capacitance and it is calculated by:

$$C_{EQ} = 2\pi\varepsilon_0 \frac{1}{\ln\left(\frac{2h}{r_{GMR}}\right)} \Rightarrow Y_{EQ} = j\omega \frac{2\pi\varepsilon_0}{\ln\left(\frac{2h}{r_{GMR}}\right)} \tag{41}$$

II.3. Alternative Method for Determination of Equivalent Conductor from a Bundled Conductor [36]

The generic bundled conductor shown in Figure II.2 is shown again in Figure II.3, considering the line sending and receiving terminals as well as a non-ideal ground. For sub-conductor 1, it is related the currents I_{A1} and I_{B1} that are associated to the terminal A and terminal B, respectively. These relations are similar to the other sub-conductors. The I_A current is the sum of all sub-conductor currents in terminal A and the I_B current is the sum of all sub-conductor currents in terminal B. It is shown in the next equation. In Figure II.3, for the all sub-conductors, the voltage level is V_A in terminal A and V_B in terminal B. Basing on this and using distributed line parameters, it is obtained the relationships among voltages and currents shown in the next equations. These equations can be related to an equivalent conductor that represents the bundled conductor.

$$V_A = V_B \cosh(\gamma \cdot d_L) - Z_C \cdot I_B \sinh(\gamma \cdot d_L) \tag{42}$$

$$I_A = \frac{V_B}{Z_C}\cosh(\gamma \cdot d_L) - I_B \sinh(\gamma \cdot d_L)$$

In the last equations, d_L is the length of the equivalent conductor that is equal to the length of each sub-conductor. The propagation function (γ) and the characteristic impedance (Z_C) are calculated using the equivalent longitudinal impedance and the equivalent shunt admittance.

$$\gamma = \sqrt{Z_{EQ} \cdot Y_{EQ}} \quad and \quad Z_C = \sqrt{\frac{Z_{EQ}}{Y_{EQ}}} \tag{43}$$

The longitudinal impedance and the shunt admittance matrices, Z_{SUB} and Y_{SUB}, respectively, which represent the characteristics of sub-conductors, are frequency dependent. In this case, there are couplings among the sub-conductors. Because of this, the elements that are out of the mentioned matrices' main diagonal are not null. The elements of Z_{SUB} and Y_{SUB} are calculated following the development shown in item II.1. If it is set up the relationships between these matrices and the equivalent conductor characteristics, it is possible to carry out the procedure proposed in this item. It is made using the modal transformation applied to the Z_{SUB} and Y_{SUB} matrices. Manipulating the obtained mode matrices, it is set up the mentioned relationships though the Z_C and γ values. In this case, the modal transformation is used because there are not mutual couplings in mode domain. In this domain, a bundled conductor with n sub-conductors is represented by n uncoupled propagation modes.

$$Z_{SUB} = \begin{bmatrix} Z_{11} & \cdots & Z_{1n} \\ \vdots & \ddots & \vdots \\ Z_{1n} & \cdots & Z_{nn} \end{bmatrix} \quad and \quad Y_{SUB} = \begin{bmatrix} Y_{11} & \cdots & Y_{1n} \\ \vdots & \ddots & \vdots \\ Y_{1n} & \cdots & Y_{nn} \end{bmatrix} \qquad (44)$$

The mode matrices are:

$$Z_{SUB\,M} = \begin{bmatrix} Z_{M1} & 0 & \cdots & 0 \\ 0 & Z_{M2} & \vdots & \vdots \\ \vdots & \vdots & \ddots & 0 \\ 0 & \cdots & 0 & Z_{Mn} \end{bmatrix} = T^T \cdot Z \cdot T \qquad (45)$$

$$Y_{SUB\,M} = \begin{bmatrix} Y_{M1} & 0 & \cdots & 0 \\ 0 & Y_{M2} & \vdots & \vdots \\ \vdots & \vdots & \ddots & 0 \\ 0 & \cdots & 0 & Y_{Mn} \end{bmatrix} = T^{-1} \cdot Y \cdot T^{-T}$$

The $Z_{SUB\,M}$ matrix is the mode longitudinal impedance one. The $Y_{SUB\,M}$ matrix is the shunt admittance one. The T matrix is the eigenvector matrix obtained from the YZ matricial product. Using the mode domain, the propagation modes are uncoupled and can be independently modeled applying the next equations where it is considered a generic propagation mode related to the analyzed bundled conductor:

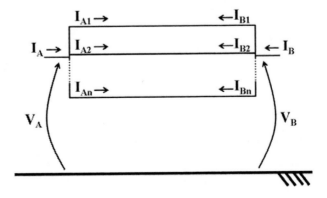

Figure II.3. Sending and receiving terminals of bundled conductor.

$$V_{MAi} = V_{MBi} \cosh(\gamma_{Mi} \cdot d_L) - Z_{CMi} \cdot I_{MBi} \sinh(\gamma_{Mi} \cdot d_L)$$

$$I_{MAi} = \frac{V_{MBi}}{Z_{CMi}} \cosh(\gamma_{Mi} \cdot d_L) - I_{MBi} \sinh(\gamma_{Mi} \cdot d_L) \quad \Bigg\}, \quad i = 1, ..., n \qquad (46)$$

In this case, Z_{CMi} and γ_{Mi} are the characteristic impedance and the propagation function of a generic propagation mode (i mode). The currents associated to this mode are I_{MAi} and I_{MBi}. The voltages are V_{MAi} and V_{MBi}. From the mode matrices, it is written:

$$\gamma_{Mi} = \sqrt{Z_{Mi} \cdot Y_{Mi}} \quad and \quad Z_{CMi} = \sqrt{\frac{Z_{Mi}}{Y_{Mi}}} \qquad (47)$$

Using matricial equations, it can be obtained the following linear system:

$$V_{MA} = \theta_1 \cdot V_{MB} - \theta_2 \cdot I_{MB} \quad and \quad I_{MA} = \theta_3 \cdot V_{MB} - \theta_4 \cdot I_{MB} \qquad (48)$$

Basing on the equations (46) and (48), it is defined:

$$V_{MA} = \begin{bmatrix} V_{MA1} \\ \vdots \\ V_{MAn} \end{bmatrix}, \quad I_{MA} = \begin{bmatrix} I_{MA1} \\ \vdots \\ I_{MAn} \end{bmatrix}, \quad V_{MB} = \begin{bmatrix} V_{MB1} \\ \vdots \\ V_{MBn} \end{bmatrix}, \quad I_{MB} = \begin{bmatrix} I_{MB1} \\ \vdots \\ I_{MBn} \end{bmatrix} \qquad (49)$$

It is also defined:

$$\theta_1(i,i) = \cosh(\gamma_{Mi} \cdot d_L)$$

$$\theta_2(i,i) = Z_{CMi} \sinh(\gamma_{Mi} \cdot d_L)$$

$$\theta_3(i,i) = \frac{\cosh(\gamma_{Mi} \cdot d_L)}{Z_{CMi}} = \frac{\theta_1(i,i)}{Z_{CMi}} \qquad (50)$$

$$\theta_4(i,i) = \sinh(\gamma_{Mi} \cdot d_L) = \frac{\theta_2(i,i)}{Z_{CMi}}$$

The θ_1, θ_2, θ_3 and θ_4 matrices are diagonal matrices where only the main diagonal elements are not null.

The vectors shown in equations (49) are related to the phase domain using the following:

$$V_{MA} = T^T \cdot V_A, \quad I_{MA} = T^{-1} \cdot I_A \qquad (51)$$

$$V_{MB} = T^T \cdot V_B, \quad I_{MB} = T^{-1} \cdot I_B$$

Using these relations, the voltages and currents are converted to the phase domain. If it is applied to the obtained linear system, it is obtained:

$$V_A = T^{-T} \cdot \theta_1 \cdot T^T \cdot V_B - T^{-T} \cdot \theta_2 \cdot T^{-1} \cdot I_B$$

$$I_A = T \cdot \theta_3 \cdot T^T \cdot V_B - T \cdot \theta_4 \cdot T^{-1} \cdot I_B$$

(52)

Manipulating the last equations, it is obtained:

$$I_B = T \cdot \theta_2^{-1} \cdot \theta_1 \cdot T^T \cdot V_B - T \cdot \theta_2^{-1} \cdot T^T \cdot V_A$$
$$\Downarrow$$
$$I_A = T \cdot \theta_4 \cdot \theta_2^{-1} \cdot T^T \cdot V_A + \left(T \cdot \theta_3 \cdot T^T - T \cdot \theta_4 \cdot \theta_2^{-1} \cdot \theta_1 \cdot T^T \right) \cdot V_B$$

(53)

Simplifying the equation related to the I_A current matrix, it is obtained:

$$I_A = A \cdot V_A + B \cdot V_B$$

(54)

In this case, A and B matrices are defined as:

$$A = T \cdot \theta_4 \cdot \theta_2^{-1} \cdot T^T \quad and \quad B = T \cdot \theta_3 \cdot T^T - T \cdot \theta_4 \cdot \theta_2^{-1} \cdot \theta_1 \cdot T^T$$

(55)

Describing the last equations in detail, it is obtained:

$$\begin{bmatrix} I_{A1} \\ \vdots \\ I_{An} \end{bmatrix} = \begin{bmatrix} A_{11} & \cdots & A_{1n} \\ \vdots & \ddots & \vdots \\ A_{n1} & \cdots & A_{nn} \end{bmatrix} \cdot \begin{bmatrix} V_A \\ \vdots \\ V_A \end{bmatrix} + \begin{bmatrix} B_{11} & \cdots & B_{1n} \\ \vdots & \ddots & \vdots \\ B_{n1} & \cdots & B_{nn} \end{bmatrix} \cdot \begin{bmatrix} V_B \\ \vdots \\ V_B \end{bmatrix}$$

(56)

The addition of the linear system equations shown in equation (56) leads to:

$$I_A = \sum_{i=1}^{n} \sum_{k=1}^{n} A_{ik} \cdot V_A + \sum_{i=1}^{n} \sum_{k=1}^{n} B_{ik} \cdot V_B$$

(57)

The linear system that describes the bundled conductor is rewritten as following:

$$V_B = V_A \cosh(\gamma \cdot d_L) - Z_C \cdot I_A \sinh(\gamma \cdot d_L)$$

$$I_B = \frac{V_A}{Z_C} \cosh(\gamma \cdot d_L) - I_A \sinh(\gamma \cdot d_L)$$

(58)

Manipulating the first equation of the shown linear system, it is obtained the I_A current as a function of only the V_A and V_B voltages:

$$I_A = \frac{\cosh(\gamma \cdot d_L)}{Z_C \cdot \sinh(\gamma \cdot d_L)} \cdot V_A - \frac{1}{Z_C \cdot \sinh(\gamma \cdot d_L)} \cdot V_B \qquad (59)$$

Equaling the equations (57) and (59), it is determined the propagation function and the characteristic impedance from the elements of the A and B matrices as following:

$$\gamma = \frac{1}{d_L} \operatorname{arc\,cosh}\left(-\frac{\sum_{i=1}^{n}\sum_{k=1}^{n} A_{ik}}{\sum_{i=1}^{n}\sum_{k=1}^{n} B_{ik}} \right) \qquad (60)$$

$$Z_C = \frac{1}{\sqrt{\left(\sum_{i=1}^{n}\sum_{k=1}^{n} A_{ik}\right)^2 - \left(\sum_{i=1}^{n}\sum_{k=1}^{n} B_{ik}\right)^2}}$$

These values can be applied to the determination of the equivalent conductor characteristics. It is shown in the next equations.

$$Z_{EQ} = \gamma \cdot Z_C = \frac{1}{d_L \cdot \sqrt{\left(\sum_{i=1}^{n}\sum_{k=1}^{n} A_{ik}\right)^2 - \left(\sum_{i=1}^{n}\sum_{k=1}^{n} B_{ik}\right)^2}} \operatorname{arc\,cosh}\left(-\frac{\sum_{i=1}^{n}\sum_{k=1}^{n} A_{ik}}{\sum_{i=1}^{n}\sum_{k=1}^{n} B_{ik}} \right) \qquad (61)$$

$$Y_{EQ} = \frac{\gamma}{Z_C} = \frac{\sqrt{\left(\sum_{i=1}^{n}\sum_{k=1}^{n} A_{ik}\right)^2 - \left(\sum_{i=1}^{n}\sum_{k=1}^{n} B_{ik}\right)^2}}{d_L} \operatorname{arc\,cosh}\left(-\frac{\sum_{i=1}^{n}\sum_{k=1}^{n} A_{ik}}{\sum_{i=1}^{n}\sum_{k=1}^{n} B_{ik}} \right)$$

Simplifying the terms, it is defined one term for the addition of all the A matrix elements and other term for the addition of all the B matrix elements:

$$S_A = \sum_{i=1}^{n}\sum_{k=1}^{n} A_{ik} \quad and \quad S_B = \sum_{i=1}^{n}\sum_{k=1}^{n} B_{ik} \qquad (62)$$

Applying these terms, the equations (59) are rewritten. The Z_{EQ} value is:

$$Z_{EQ} = \frac{1}{d_L \cdot \sqrt{S_A^2 - S_B^2}} \operatorname{arc\,cosh}\left(-\frac{S_A}{S_B} \right) \qquad (63)$$

The Y_{EQ} value is:

$$Y_{EQ} = \frac{\sqrt{S_A^2 - S_B^2}}{d_L} \operatorname{arc cosh}\left(-\frac{S_A}{S_B}\right) \qquad (64)$$

From these obtained equivalent values, it is determined the conductor characteristics that is equivalent to a bundled conductor. Using the Z_{EQ} and Y_{EQ} values, the R_{EQ}, L_{EQ} and C_{EQ} values can be obtained for a determined frequency range, defining the equivalent conductor for the analyzed bundled conductor.

As a sample for applying the proposed procedure, it is shown in the next figure a non-conventional bundled conductor that is composed by 7 sub-conductors. There is a central conductor with a 3.5 cm radius. The radius of each surrounding sub-conductor is 1.5 cm. The ground resistivity is considered as 1000 Ω·m. The height related to the geometrical center of this bundled conductor is 12 m and it corresponds to the height of the central sub-conductor.

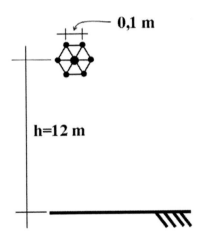

Figure II.4. Non-conventional bundled conductor

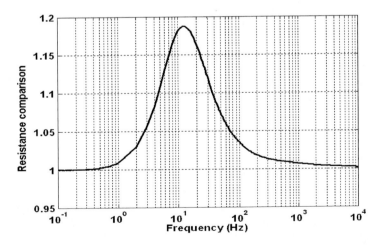

Figure II.5. Comparisons of resistance values calculated from the GMR method and the proposed procedure for the equivalent conductor

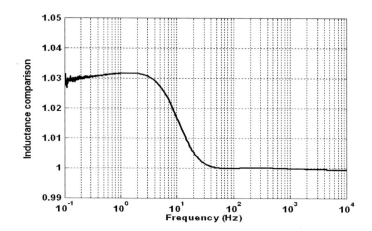

Figure II.6. Comparisons of inductance values calculated from the GMR method and the proposed procedure for the equivalent conductor

The parameters of the equivalent conductor that represents the shown bundled conductor are calculated using the GMR method and the procedure proposed in this item. After this, the results are compared and shown in Figures II.5 and II.6. In this case, Figure II.5 show the comparisons obtained for resistance values. In the frequency range from 10 Hz to 100 Hz, the difference between the two used methods reaches the highest value that is about 18 %. Then, the unbalanced distribution of currents leads to a great difference in determination of the resistance value for the steady state frequency range. It is also concluded for the comparisons results of inductance values shown in Figure II.6. From low frequencies to 5 Hz, there is approximately a constant difference between the both methods. This difference is about 3 %.

For bundled conductors with balanced current distribution, the proposed procedure and the GMR method get equal results. The GMR method, basing on the results of this item, it is not adequate, if there is unbalanced current distribution on the analyzed bundle conductor. Then, the proposed procedure is a useful numeric tool, because it is adequate for calculating the equivalent conductor of bundled conductors when these have balanced and unbalanced current distribution.

III. UNTRANSPOSED SYMMETRICAL THREE-PHASE TRANSMISSION LINE MODAL REPRESENTATION USING TWO TRANSFORMATION MATRICES [37, 38]

It is proposed in this section a modal representation for untransposed symmetrical three-phase transmission lines basing on the application of two transformation matrices. In this case, it is considered transmission lines that have a vertical symmetry plane. The first applied transformation matrix is Clarke's matrix. This matrix decomposes the line phases in one exact mode and two non-exact modes, called quasi-modes. From these quasi-modes, it is obtained the other two exact modes applying a 2-order modal transformation matrix. The direct application of modal transformation for three-phase transmission lines uses 3-order frequency dependent transformation matrices. So, the advantage of the proposed way for modal transformation application is the reduction of the frequency dependent elements that compose this transformation. In this case, Clarke's matrix is frequency independent and it is necessary only the 2-order modal transformation matrix application.

For electromagnetic transient analyses in power systems, it can be considered numeric methods based on the phase domain or modal domain. Applying the modal domain, usually, it is considered a hybrid model where it is included the interactions between the phase domain and the mode domain using the modal transformations. The wave propagation on the line is determined in mode domain and the interactions with the phase domain are necessary for including the obtained results in the remaining network power system.

When there is a vertical symmetry plane in three-phase transmission lines and these lines are not transposed, the Clarke's matrix application obtains the α, β and 0 components. The β component is an exact mode because it does not have couplings with the other both components. Between the other both components, α and 0, there is a coupling and they are not be considered exact modes. Because of this, some authors have called these components as the α and 0 quasi-modes. For some cases, this coupling can be neglected and the α and 0 exact modes can be changed into the α and 0 quasi-modes, respectively. For general symmetrical three-phase transmission line cases, the α and 0 quasi-modes can be treated as a two-phase transmission line and this hypothetical line is changed into two uncoupled two exact modes.

III.1. Obtaining the β Exact Mode Applying Clarke's Matrix

Considering an unstransposed three-phase transmission line with a vertical symmetry plane, it is used the next figure schematizing this situation. The central phase height is different of the height of adjacent phases. So, the heights of adjacent phases are equal. The horizontal distance between each adjacent phase and the central phase, d_{HD} in the next figure, is also equal.

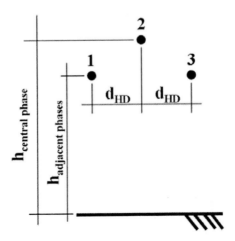

Figure III.1. A symmetrical three-phase transmission line schema

For the transmission line schematize in the last figure, the Z and Y matrices can be described as following:

$$Z_S = \begin{bmatrix} Z_{11} & Z_{12} & Z_{13} \\ Z_{12} & Z_{22} & Z_{12} \\ Z_{13} & Z_{12} & Z_{11} \end{bmatrix} \quad and \quad Y_S = \begin{bmatrix} Y_{11} & Y_{12} & Y_{13} \\ Y_{12} & Y_{22} & Y_{12} \\ Y_{13} & Y_{12} & Y_{11} \end{bmatrix} \tag{65}$$

Because the symmetry plane, the characteristics related to the phase 1 of Figure III.1 are equal to those related to the phase 3 of the same figure. It leads to the results shown in the last equation. Applying Clarke's matrix to the Z_S and Y_S matrices, it is obtained these matrices in the new domain. It is shown in equations (66). This transformation is not the exact modal transformation because Clarke's matrix is not an eigenvector matrix for the considered case. Completing the modal transformation, it is only considered the α and 0 quasi-modes. It is used this denomination because there is coupling between these two components.

$$Z_{MS} = T_{CL}^T \cdot Z_{SL} \cdot T_{CL} = \begin{bmatrix} Z_{Q\alpha} & 0 & Z_{\alpha 0} \\ 0 & Z_\beta & 0 \\ Z_{\alpha 0} & 0 & Z_{Q0} \end{bmatrix} \tag{66}$$

$$Y_{MS} = T_{CL}^T \cdot Y_{SL} \cdot T_T = \begin{bmatrix} Y_{Q\alpha} & 0 & Y_{\alpha 0} \\ 0 & Y_\beta & 0 \\ Y_{\alpha 0} & 0 & Y_{Q0} \end{bmatrix}$$

The structure of Clarke's matrix is:

$$T_{CL} = \begin{bmatrix} -\frac{1}{\sqrt{6}} & \frac{1}{\sqrt{2}} & \frac{1}{\sqrt{3}} \\ \frac{2}{\sqrt{6}} & 0 & \frac{1}{\sqrt{3}} \\ -\frac{1}{\sqrt{6}} & -\frac{1}{\sqrt{2}} & \frac{1}{\sqrt{3}} \end{bmatrix} \tag{67}$$

The β component obtained from the Clarke's matrix application is an exact mode because the $Z_{\alpha\beta}$, $Z_{\beta 0}$, $Y_{\alpha\beta}$, $Y_{\beta 0}$ couplings in equations (66) are null. Separating the coupled quasi-modes, it is obtained:

$$Z_{Q\alpha 0} = \begin{bmatrix} Z_{Q\alpha} & Z_{\alpha 0} \\ Z_{\alpha 0} & Z_{Q0} \end{bmatrix} \quad and \quad Y_{Q\alpha 0} = \begin{bmatrix} Y_{Q\alpha} & Y_{\alpha 0} \\ Y_{\alpha 0} & Y_{Q0} \end{bmatrix} \tag{68}$$

III.2. Obtaining the α and 0 Exact Modes Applying a 2-Order Transformation Matrix

The $Z_{Q\alpha 0}$ and $Y_{Q\alpha 0}$ can be represented as a hypothetic two-phase transmission line that has not a symmetry plane. This hypothetic line is shown in the next figure and the angle between the both phases ($\theta_{\alpha 0}$) should be different of 0 or 180 °. In case of these values, the couplings between α and 0 quasi-modes are null.

From the $Z_{Q\alpha 0}$ and $Y_{Q\alpha 0}$ matrices, it is obtaining the eigenvector matrix of the $Y_{Q\alpha 0}Z_{Q\alpha 0}$ matricial product identified as the $T_{\alpha 0}$ transformation matrix. Applying this matrix, it is obtaining the exact α and 0 modes as showing in the following:

$$Z_{M\alpha 0} = T_{\alpha 0}^{-1} \cdot Z_{Q\alpha 0} \cdot T_{\alpha 0} = \begin{bmatrix} Z_\alpha & 0 \\ 0 & Z_0 \end{bmatrix}$$

$$Y_{M\alpha 0} = T_{\alpha 0}^{-1} \cdot Y_{Q\alpha 0} \cdot T_{\alpha 0} = \begin{bmatrix} Y_\alpha & 0 \\ 0 & Y_0 \end{bmatrix} \tag{69}$$

The $T_{\alpha 0}$ matrix structure is:

$$T_{\alpha 0} = \begin{bmatrix} T_{\alpha 0\ 11} & T_{\alpha 0\ 21} \\ T_{\alpha 0\ 12} & T_{\alpha 0\ 22} \end{bmatrix} \tag{70}$$

Recomposing the three-order transformation matrix, it is considered the $T_{\alpha 0}$ matrix elements as following:

$$T_{3\alpha 0} = \begin{bmatrix} T_{\alpha 0\ 11} & 0 & T_{\alpha 0\ 21} \\ 0 & 1 & 0 \\ T_{\alpha 0\ 12} & 0 & T_{\alpha 0\ 22} \end{bmatrix} \tag{71}$$

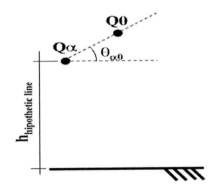

Figure III.2. Hypothetic line for the coupled α and 0 quasi-mode components

The modal transformation matrix for the symmetrical three-phase transmission line considering the untransposed case is obtaining as:

$$T_{SL} = T_{CL} \cdot T_{3\alpha 0} = \begin{bmatrix} \left(\dfrac{-T_{\alpha 0\,11}}{\sqrt{6}} + \dfrac{T_{\alpha 0\,12}}{\sqrt{3}}\right) & \dfrac{1}{\sqrt{2}} & \left(\dfrac{-T_{\alpha 0\,21}}{\sqrt{6}} + \dfrac{T_{\alpha 0\,22}}{\sqrt{3}}\right) \\ \left(\dfrac{2 \cdot T_{\alpha 0\,11}}{\sqrt{6}} + \dfrac{T_{\alpha 0\,12}}{\sqrt{3}}\right) & 0 & \left(\dfrac{2 \cdot T_{\alpha 0\,11}}{\sqrt{6}} + \dfrac{T_{\alpha 0\,12}}{\sqrt{3}}\right) \\ \left(\dfrac{-T_{\alpha 0\,11}}{\sqrt{6}} + \dfrac{T_{\alpha 0\,12}}{\sqrt{3}}\right) & \dfrac{-1}{\sqrt{2}} & \left(\dfrac{-T_{\alpha 0\,21}}{\sqrt{6}} + \dfrac{T_{\alpha 0\,22}}{\sqrt{3}}\right) \end{bmatrix} \qquad (72)$$

For symmetrical three-phase transmission lines, using the proposed procedure, the modal transformation matrix is obtained from the matricial product between Clarke's matrix and the $T_{3\alpha 0}$ one. This last matrix is calculated based on the α and 0 quasi-modes. After the Clarke's matrix application, these quasi-modes have coupled yet and it is necessary manipulated then for obtaining the exact α and 0 modes. For this manipulation, it is obtained the elements of the $T_{3\alpha 0}$ matrix that completes the modal transformation for the analyzed case.

The advantage of the proposed procedure is the determination of only four frequency dependent elements that composed the modal transformation matrix. If it is used a procedure adequate to a general three-phase transmission line case, it is necessary all nine elements of the modal transformation matrix that depends on the frequency.

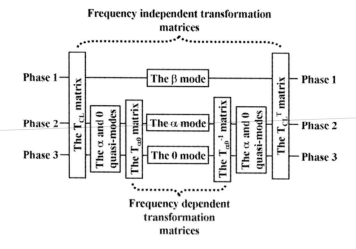

Figure III.3. Schema for applying the procedure proposed in this section

The schema shown in Figure III.3 illustrates the procedure proposed in this section. Applying this schematized procedure for electromagnetic transient and wave propagation simulation can imply in convolution numeric method uses. The convolution numeric methods are necessary for solving the interactions between the phase domain and mode domain through the frequency transformation matrices. So, the proposed procedure can be an alternative numeric method for changing the phase domain values into mode domain values. Then, an actual three-phase line that belongs to Brazilian's utilities, is a sample for applying the proposed procedure and shown in Figure III.4.

III.3. An Actual Transmission Line Sample

About the characteristics of the line shown in the next figure, it is a 440 kV transmission line with a vertical symmetry plane. The central phase conductor height is 27.67 m on the tower. The height of adjacent phase conductors is 24.07 m. Every phase is composed of four sub-conductors distributed in a square shape with 0.4 m side length. Every sub-conductor is an ACSR type one (ACSR 26/7 636 MCM) with internal diameter of 0.93 cm and a external diameter of the 2.52 cm. The phase sub-conductor resistivity is 0.089899 Ω/km and the sag at the midspan is 13.43 m. The earth resistivity is considered constant (1000 Ω.m). There are two ground wires and they are EHS 3/8" with the resistivity of 4.188042 Ω/km. The diameters of these cables are 0.9144 cm. The height of these cables on the tower is 36.00 m. The sag of the ground wires at the midspan is 6.40 m. Each ground wire is composed by a single conductor.

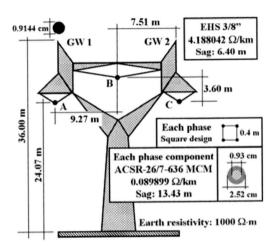

Figure III.4. An actual transmission line

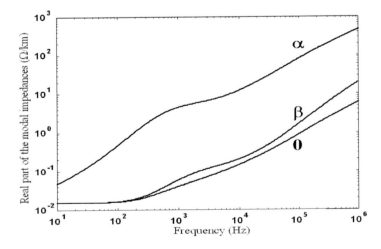

Figure III.5. Mode resistances of the actual transmission line used as a sample

The last figure shows the real part of mode impedances calculated using the procedure proposed in this item.

In the next figure, it is shown the results obtaining for mode inductances that are the imaginary part of the mode impedances. In this case, it is also used the proposed procedure.

For checking the efficiency and the accuracy of the proposed procedure, it is carried out a frequency scan testing mode domain using a 1 pu voltage step source linked to the phase 1 of the shown actual transmission line. The sending terminals of the other both phases are in short-circuit. The receiving terminals of all phases are linked to the impedances which values are equal to the line characteristic impedance (Z_C). It avoids the propagated wave reflections. Applying the proposed procedure, it is calculated the modes of the actual transmission line sample and the 1 pu voltage source is inserted. This is schematized in Figure III.7.

The obtained results of the mentioned test are compared to those obtained from the 3-order frequency dependent modal transformation matrix application. This matrix is calculated using Newton-Raphson's numeric method. Both obtained results are compared to those related to Clarke's matrix. Figure III.8 shows these comparisons that are related to the voltage values that depend on the frequency for the phase 2 receiving terminal.

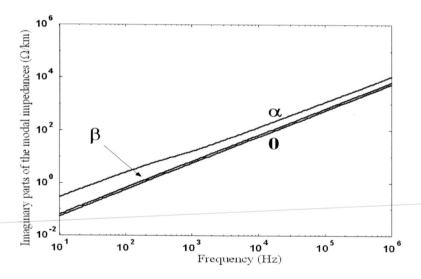

Figure III.6. Mode inductances of the actual transmission line used as a sample.

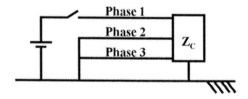

Figure III.7. Energization of the actual transmission line phase 1.

Based on the comparisons shown in the next figure, the proposed procedure can be classified an exact numeric method for calculating the modal transformation of a three-phase transmission line with a vertical symmetry plane. Besides this, it is observed that there are differences among the results related to Clarke's matrix and the exact results obtained from the both exact methods until 10 kHz. Discussing if these differences are significant or not, it is analyzed in the next items the application of Clarke's matrix for electromagnetic transient simulations considering untransposed transmission lines.

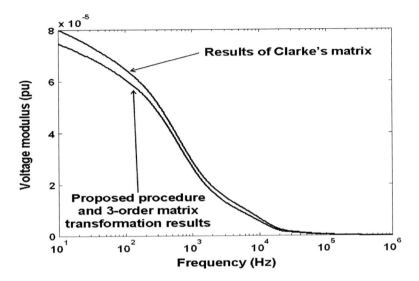

Figure III.8. Comparisons of frequency scan test results in mode domain

IV. SINGLE REAL TRANSFORMATION MATRIX APPLICATIONS FOR UNTRANSPOSED THREE-PHASE TRANSMISSION LINES [39-41]

Analyzing transposed three-phase transmission lines, single real matrices can be carried out exact modal transformations because the mentioned matrices are eigenvector ones in these cases. For untransposed three-phase transmission lines, if single real matrices are used as transformation ones, it is not obtained exact modal transformations. However, single real transformation matrices can be good approximations for the exact transformation matrices and it is analyzed in this item considering three-phase lines.

Considering transmission line analyses, the frequency influence is significant for line longitudinal parameters. Because of this, the line representation in time domain becomes complex. There are models where it is applied phase mode or modal transformation, considering the modeling in mode domain and improving the frequency dependent line parameter representation. For this, it is necessary to take to frequency dependent transformation matrices, because the line characteristics depend on the frequency. Frequency independent transformation matrices can be an alternative for this analysis type, if they lead to good approximations for the exact transformations. The interest about this alternative is based on these characteristics of the mentioned constant matrix application: single, frequency independent, line parameter independent. The main explanation for the interest in the frequency independent matrix application as a modal transformation matrix is that, considering transposed transmission line, the modal transformation matrices are frequency independent and, the most of these cases, they are also line parameter independent.

There are three main matrices that have used as modal transformation matrices for three-phase transmission: Clarke's matrix, Fortescue's one and Karrenbauer's one. These matrices are exact eigenvector and modal transformation matrices for transposed three-phase transmission lines. For three-phase systems, Fortescue's matrix is degenerated into symmetrical component one. Clarke's matrix is a linear combination of the symmetrical component matrix, searching for real matrix elements. So, Clarke's matrix, besides the mentioned characteristics, incorporated more two characteristics into the application analyzed in this item: this matrix is real and its application is identical for voltage and current manipulating. With Clarke's matrix use, mathematical model simplifications are obtained and the developed model can be applied directly in time domain programs without numeric convolution methods procedures.

In this section, for typical untransposed three-phase transmission lines, the eigenvector and eigenvalue analyses are carried out using Clarke's matrix as an alternative for modal transformation matrix. It is considered symmetrical and asymmetrical three-phase transmission line cases. When the Clarke's matrix results, called quasi-modes, are compared to the eigenvalues, the relative errors are reasonably small and can be considered negligible as well as accepted for most applications.

IV.1. Mathematical Modeling for Modal Transformation Application

When transmission lines are submitted to electromagnetic transient phenomena, the relations among transversal voltage vector (V_{PH}) and longitudinal current one (I_{PH}) in phase domain can be described by the following equations. As previously mentioned, the Z and Y matrices are the longitudinal impedance and shunt admittance ones, respectively.

$$-\frac{dV_{PH}}{dx} = Z \cdot I_{PH} \quad and \quad -\frac{dI_{PH}}{dx} = Y \cdot V_{PH} \tag{73}$$

In mode domain, the V_{PH} and I_{PH} vectors are changed into the mode voltage (V_M) and the mode current (I_M) vectors, respectively, using the following:

$$V_{PH} = T_V \cdot V_M \quad and \quad I_{PH} = T_I \cdot I_M \tag{74}$$

The T_I matrix is obtained from equation (2). The T_V matrix is obtained from the next equation.

$$T_V^{-1} \cdot Z \cdot Y \cdot T_V = \lambda \tag{75}$$

Equaling and manipulating equations (2) and (75), it is obtained:

$$T_V^{-1} = T_I^T \tag{76}$$

Applying these results on equations (73), it is obtained:

$$-\frac{d(T_V \cdot V_M)}{dx} = Z \cdot T_I \cdot I_M \quad and \quad -\frac{d(T_I \cdot I_M)}{dx} = Y \cdot T_V \cdot V_M$$

$$\Downarrow \qquad\qquad (77)$$

$$-\frac{dV_M}{dx} = T_I^T \cdot Z \cdot T_I \cdot I_M \quad and \quad -\frac{dI_M}{dx} = T_I^{-1} \cdot Y \cdot T_I^{-T} \cdot V_M$$

And these last equations also lead to the relations shown in equations (4). In this item, for three-phase transmission line cases, equations (4) are modified changing the exact eigenvector matrices into Clarke's matrix as described following:

$$T_V = T_{CL} \quad and \quad T_I = T_{CL} \qquad (78)$$

Clarke's inverse matrix is Clarke's transposed matrix:

$$T_{CL}^{-1} = T_{CL}^T \qquad (79)$$

With the mentioned changes, it is obtained:

$$Z_{MCL} = T_{CL}^T \cdot Z \cdot T_{CL} \quad and \quad Y_{MCL} = T_{CL}^T \cdot Y \cdot T_{CL} \qquad (80)$$

It is also obtained:

$$\lambda_{ICL} = T_{CL}^T \cdot Y \cdot Z \cdot T_{CL} \quad and \quad \lambda_{VCL} = T_{CL}^T \cdot Z \cdot Y \cdot T_{CL} \qquad (81)$$

For general three-phase transmission line cases, the λ_{ICL} and λ_{VCL} matrices are not equal because the ZY and YZ matricial products are neither equal. Changing the eigenvector matrices into Clarke's matrix, the modal transformation is not exact, the quasi-modes are obtained and the previously mentioned matrices, Z_{MCL}, Y_{MCL}, λ_{ICL} as well as λ_{VCL}, are not diagonal ones.

For transposed three-transmission line cases, considering ideal assumptions, the YZ and ZY matricial products become equal, becoming equal the Z_{MCL} and Y_{MCL} matrices to the exact correspondent ones as well as the λ_{ICL} and λ_{VCL} matrices to the λ matrix. In this case, it is obtained the following equations where the T letter addition in subscript term identifies the transposed cases:

$$Z_{TM} = T_{CL}^T \cdot Z \cdot T_{CL} = Z_M = T_I^T \cdot Z \cdot T_I = \begin{bmatrix} Z_\alpha & 0 & 0 \\ 0 & Z_\beta & 0 \\ 0 & 0 & Z_0 \end{bmatrix} \qquad (82)$$

$$Y_{TM} = T_{CL}^T \cdot Y \cdot T_{CL} = Z_M = T_I^T \cdot Y \cdot T_I = \begin{bmatrix} Y_\alpha & 0 & 0 \\ 0 & Y_\beta & 0 \\ 0 & 0 & Y_0 \end{bmatrix}$$

For the λ_{ICL} and λ_{VCL} matrices, the equations are changed into:

$$\lambda_{TI} = T_{CL}^{T} \cdot Y \cdot Z \cdot T_{CL} = \lambda_{TV} = T_{CL}^{T} \cdot Z \cdot Y \cdot T_{CL} = \lambda = \begin{bmatrix} \lambda_\alpha & 0 & 0 \\ 0 & \lambda_\beta & 0 \\ 0 & 0 & \lambda_0 \end{bmatrix} \quad (83)$$

In sequence, it is analyzed the application of Clarke's matrix to typical untransposed three-phase transmission line cases.

IV.2. Untransposed Three-Phase Lines with a Vertical Symmetry Plane

This transmission line type can be represented by schema shown in Figure III.1. Considering the couplings among the phases and the vertical symmetrical plane, the longitudinal impedance and shunt admittance matrices in phase domain are described by equations (65). Applying Clarke's matrix, the same matrices in mode domain are described by equations (66). The quasi-mode matrices are described as following:

$$\lambda_{IS} = \begin{bmatrix} \lambda_{IS\alpha} & 0 & \lambda_{IS\alpha 0} \\ 0 & \lambda_\beta & 0 \\ \lambda_{IS\alpha 0} & 0 & \lambda_{IS0} \end{bmatrix} \quad and \quad \lambda_{VS} = \begin{bmatrix} \lambda_{VS\alpha} & 0 & \lambda_{VS\alpha 0} \\ 0 & \lambda_\beta & 0 \\ \lambda_{VS\alpha 0} & 0 & \lambda_{VS0} \end{bmatrix} \quad (84)$$

The structure of the λ matrix is:

$$\lambda = T_I^{-1} \cdot Y \cdot Z \cdot T_I = T_I^{T} \cdot Z \cdot Y \cdot T_I^{-T} = \begin{bmatrix} \lambda_\alpha & 0 & 0 \\ 0 & \lambda_\beta & 0 \\ 0 & 0 & \lambda_0 \end{bmatrix} \quad (85)$$

Comparing the λ_{IS} and λ_{VS} matrix main diagonal elements to the correspondent λ matrix elements, it is used relative errors that are calculated by:

$$\varepsilon_K (\%) = \frac{\left(\lambda_{K\,QUASI-MODE} - \lambda_K \right)}{\lambda_K} \cdot 100, \quad K = \alpha, \beta \text{ and } 0 \quad (86)$$

In this case, the $\lambda_{K\,QUASI-MODE}$ values represent the main diagonal elements of the λ_{IS} matrix or the λ_{VS} one. The λ_K values are the non null elements of the λ matrix. The obtained relative errors are shown in the next both figures.

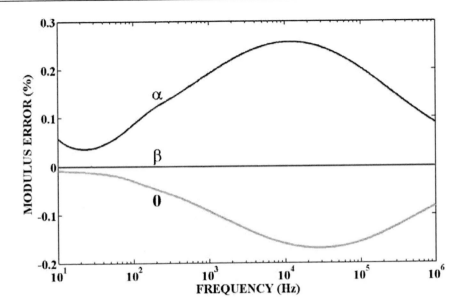

Figure IV.1. Comparisons among λ_{IS} quasi-modes and λ eigenvalues for the actual symmetrical three-phase line

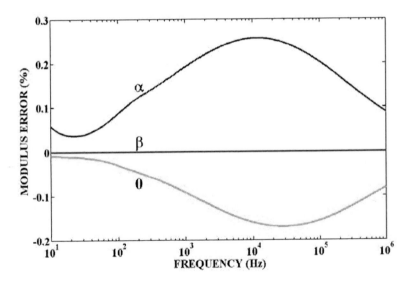

Figure IV.2. Comparisons among λ_{VS} quasi-modes and λ eigenvalues for the actual symmetrical three-phase line

Based on the relative errors shown in the last both figures, it concluded that the λ_{IS} quasi-modes are equal to the λ_{VS} ones. It can also be considered that both λ_{IS} and λ_{VS} quasi-modes are practically equal to the eigenvalues, because the relative errors can be considered negligible.

The next both figures show results related to the Z and Y matrices.

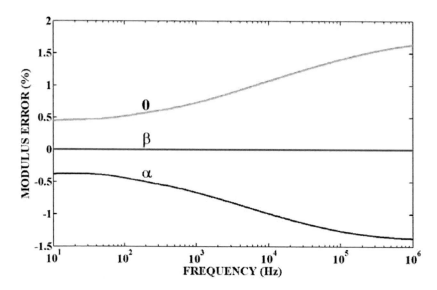

Figure IV.3. Relative errors related to the mode longitudinal impedances for the actual symmetrical three-phase transmission line

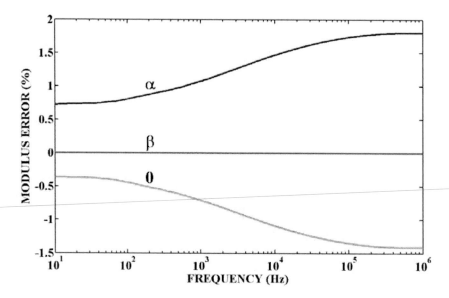

Figure IV.4. Relative errors related to the mode longitudinal admittances for the actual symmetrical three-phase transmission line

Comparing the errors shown in last both figures to those related to the eigenvalues, these errors are about 10 times higher than the quasi-mode errors. For the relative error range, these values could be considered good approximations for the exact values yet. So, in this error range, equations (66) could be rewritten as:

$$Z_{MS} = T_{CL}^T \cdot Z_{SL} \cdot T_{CL} = \begin{bmatrix} Z_\alpha & 0 & Z_{\alpha 0} \\ 0 & Z_\beta & 0 \\ Z_{\alpha 0} & 0 & Z_0 \end{bmatrix} \tag{87}$$

$$Y_{MS} = T_{CL}^T \cdot Y_{SL} \cdot T_T = \begin{bmatrix} Y_\alpha & 0 & Y_{\alpha 0} \\ 0 & Y_\beta & 0 \\ Y_{\alpha 0} & 0 & Y_0 \end{bmatrix}$$

Equations (84) are rewritten as:

$$\lambda_{IS} = \begin{bmatrix} \lambda_\alpha & 0 & \lambda_{IS\alpha 0} \\ 0 & \lambda_\beta & 0 \\ \lambda_{IS\alpha 0} & 0 & \lambda_0 \end{bmatrix} \quad and \quad \lambda_{VS} = \begin{bmatrix} \lambda_\alpha & 0 & \lambda_{VS\alpha 0} \\ 0 & \lambda_\beta & 0 \\ \lambda_{VS\alpha 0} & 0 & \lambda_0 \end{bmatrix} \tag{88}$$

Completing the analyses carried out in this item, it is calculated the relative values of the off-diagonal elements of the λ_{IS} and λ_{VS} quasi-mode matrices. For this, it is used the next equations which results are shown in the next two figures.

$$\varepsilon_{I\alpha 0}(\%) = \frac{\lambda_{I\alpha 0}}{\lambda_\alpha \ or \ \lambda_0} \cdot 100 \quad and \quad \varepsilon_{V\alpha 0}(\%) = \frac{\lambda_{V\alpha 0}}{\lambda_\alpha \ or \ \lambda_0} \cdot 100 \tag{89}$$

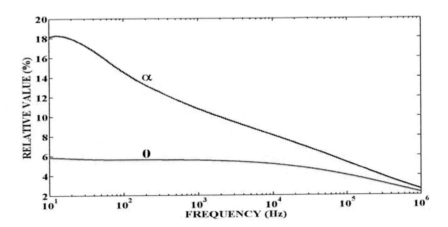

Figure IV.5. The $\lambda_{I\alpha 0}$ relative values for the actual symmetrical three-phase transmission line

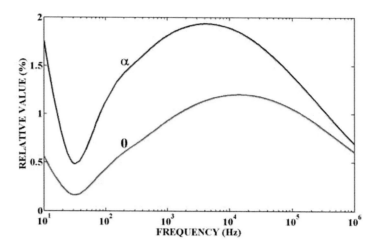

Figure IV.6. The $\lambda_{V\alpha 0}$ relative values for the actual symmetrical three-phase transmission line

In Figure IV.5, it is shown the comparisons between the $\lambda_{I\alpha 0}$ element and the λ eigenvalues that correspond to the λ_I coupled quasi-modes. Considering the $\lambda_{V\alpha 0}$ element, the comparisons to the correspondent λ_V eigenvalues are shown in Figure IV.6. These results are lower than those of Figure IV.5. So, it is concluded that the T_{CL} matrix is a good approximation for the T_V matrix, considering the actual symmetrical three-phase transmission line. It is also concluded that the main problems related to the change of the eigenvector matrices into Clarke's matrix are related to the T_I eigenvector change, because the $\lambda_{I\alpha 0}$ element reaches about 18 % of the λ_{α} eigenvalue. The influence of this coupling on electromagnetic transient analyses and simulations should be investigated. It is going carried out in the next specific item in which it is analyzed an asymmetrical three-phase line with triangular phase distribution.

Detailing these analyzes, the high relative value of the $\lambda_{I\alpha 0}$ element can be lead to high errors, if this change is applied to calculating values such as the propagation function (γ) and characteristic impedance (Z_C). For the mentioned application, relative values can reach about 130 % when compared to the coupled quasi-mode values. Because of these results, it is also investigated the application of a correction procedure for Clarke's matrix. Before this correction procedure, the results of the application of Clarke's matrix to the untransposed asymmetrical three-phase transmission lines are presented. It is shown two typical asymmetrical three-phase transmission line types.

IV.2. Untransposed Three-Phase Lines with Phase Conductors Vertically Lined

Considering the line in the next figure and applying Clarke's matrix as modal transformation matrix, there are couplings between all three quasi-modes that compose the λ_{IR} and λ_{VR} quasi-mode matrices.

The mentioned matrices are described as:

$$\lambda_{IR} = \begin{bmatrix} \lambda_{IR\alpha} & \lambda_{IR\alpha\beta} & \lambda_{IR\alpha0} \\ \lambda_{IR\alpha\beta} & \lambda_{IR\beta} & \lambda_{IR\beta0} \\ \lambda_{IR\alpha0} & \lambda_{IR\beta0} & \lambda_{IR0} \end{bmatrix} \quad and \quad \lambda_{VR} = \begin{bmatrix} \lambda_{VR\alpha} & \lambda_{VR\alpha\beta} & \lambda_{VR\alpha0} \\ \lambda_{VR\alpha\beta} & \lambda_{VR\beta} & \lambda_{VR\beta0} \\ \lambda_{VR\alpha0} & \lambda_{VR\beta0} & \lambda_{VR0} \end{bmatrix} \quad (90)$$

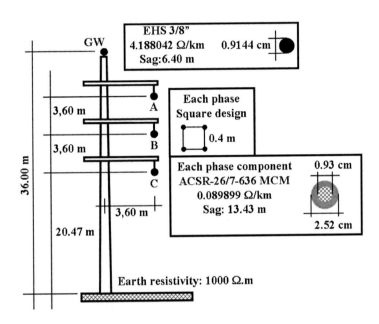

Figure IV.7. Actual asymmetrical vertical three-phase transmission line

The quasi-mode longitudinal impedance and the quasi-mode shunt admittance matrices have the following structure:

$$Z_{MR} = \begin{bmatrix} Z_{R\alpha} & Z_{\alpha\beta} & Z_{\alpha0} \\ Z_{\alpha\beta} & Z_{R\beta} & Z_{\beta0} \\ Z_{\alpha0} & Z_{\beta0} & Z_{R0} \end{bmatrix} \quad and \quad Y_{MR} = \begin{bmatrix} Y_{R\alpha} & Y_{\alpha\beta} & Y_{\alpha0} \\ Y_{\alpha\beta} & Y_{R\beta} & Y_{\beta0} \\ Y_{\alpha0} & Y_{\beta0} & Y_{R0} \end{bmatrix} \quad (91)$$

It is calculated the relative errors of the λ_{IR} and λ_{VR} quasi-modes when compared to the λ eigenvalues using equations (86). These comparisons are shown in Figures IV.8 and IV.9 and the relative errors can be considered negligible. Because of this, the last equations related to the λ_{IR} and λ_{VR} can be rewritten equaling the quasi-modes to the exact eigenvalues and maintaining the coupling among the quasi-modes. This is shown in next equations.

$$\lambda_{IR} = \begin{bmatrix} \lambda_{\alpha} & \lambda_{IR\alpha\beta} & \lambda_{IR\alpha0} \\ \lambda_{IR\alpha\beta} & \lambda_{\beta} & \lambda_{IR\beta0} \\ \lambda_{IR\alpha0} & \lambda_{IR\beta0} & \lambda_{0} \end{bmatrix} \quad and \quad \lambda_{VR} = \begin{bmatrix} \lambda_{\alpha} & \lambda_{VR\alpha\beta} & \lambda_{VR\alpha0} \\ \lambda_{VR\alpha\beta} & \lambda_{\beta} & \lambda_{VR\beta0} \\ \lambda_{VR\alpha0} & \lambda_{VR\beta0} & \lambda_{0} \end{bmatrix} \quad (92)$$

The rewritten equations are based on Figures IV.10 and IV.11.

These both figures are related to the longitudinal impedance and shunt admittance results, respectively

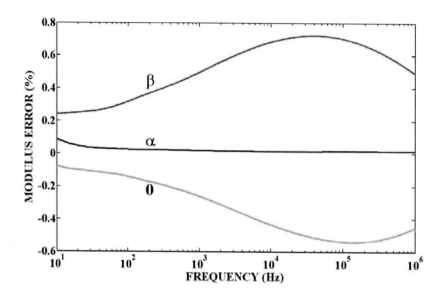

Figure IV.8. Comparisons among λ_{IR} quasi-modes and λ eigenvalues for the actual asymmetrical vertical three-phase transmission line

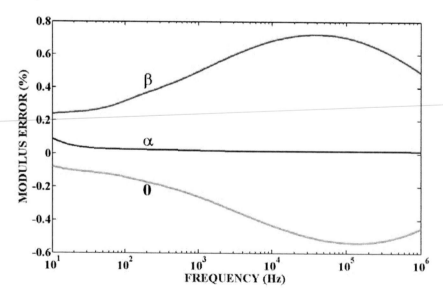

Figure IV.9. Comparisons among λ_{VR} quasi-modes and λ eigenvalues for the actual asymmetrical vertical three-phase transmission line

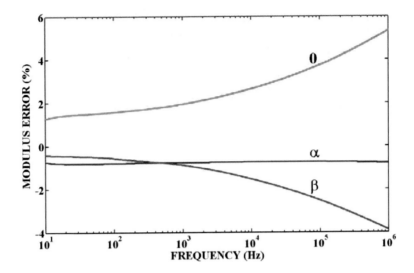

Figure IV.10. Relative errors related to the mode longitudinal impedances for the actual asymmetrical vertical three-phase transmission line

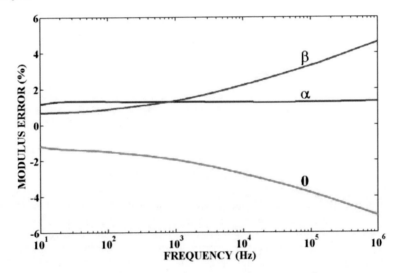

Figure IV.11. Relative errors related to the mode shunt admittances for the actual asymmetrical vertical three-phase transmission line

The errors of the quasi-mode longitudinal impedances and the quasi-mode shunt admittances reach the highest modulus values in the end of the analyzed frequency range. Depending on the application and the considered frequency, equations (91) can also be rewritten and it is obtained the following:

$$
Z_{MR} = \begin{bmatrix} Z_\alpha & Z_{\alpha\beta} & Z_{\alpha 0} \\ Z_{\alpha\beta} & Z_\beta & Z_{\beta 0} \\ Z_{\alpha 0} & Z_{\beta 0} & Z_0 \end{bmatrix} \quad and \quad Y_{MR} = \begin{bmatrix} Y_\alpha & Y_{\alpha\beta} & Y_{\alpha 0} \\ Y_{\alpha\beta} & Y_\beta & Y_{\beta 0} \\ Y_{\alpha 0} & Y_{\beta 0} & Y_0 \end{bmatrix} \tag{93}
$$

It is analyzed the coupling among the quasi-modes. In this case, the next both figures show the $\lambda_{IR\alpha\beta}$ and $\lambda_{VR\alpha\beta}$ in function of frequency values. The $\lambda_{IR\alpha\beta}$ peak value is close to 5 kHz and the $\lambda_{VR\alpha\beta}$ is in the initial of the frequency range. The results related to the T_I eigenvector change are higher than those related to the T_V eigenvector change. So, the $\lambda_{IR\alpha\beta}$ values are in the range from 0.25 % to 0.65 % while the $\lambda_{VR\alpha\beta}$ values are in the range from 0.04 % to 0.18 %. This characteristic is true for the other both coupling elements which relative values are shown in Figures IV.14, IV.15, IV.16 and IV.17. In case of Figure IV.14, the peak value is related to frequencies close to 20 Hz and it is higher than that related to the $\alpha0$ coupling of the symmetrical three-phase transmission line analyzed in the previous item. For Figure IV.15, this peak is related to initial values of the frequency range. The T_I eigenvector change shows higher relative values than the T_V eigenvector change. For the T_I eigenvector change, it is in the range from 0 to 25 % while, for T_V eigenvector change, it is in the range from 0 to 2.5 %. So, the $\lambda_{IR\alpha0}$ relative values are about 10 times higher than the $\lambda_{VR\alpha0}$ ones.

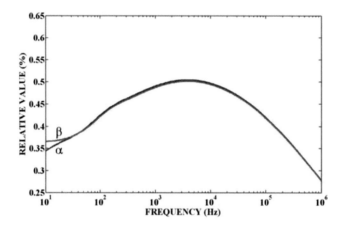

Figure IV.12. The $\lambda_{IR\alpha\beta}$ relative values for the actual asymmetrical vertical three-phase transmission line

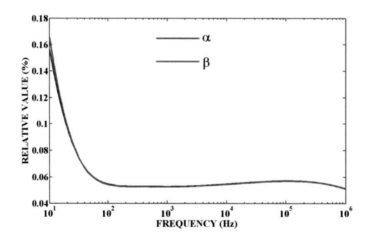

Figure IV.13. The $\lambda_{VR\alpha\beta}$ relative values for the actual asymmetrical vertical three-phase transmission line

The relative values shown in Figure IV.14 are not negligible values mainly when it is considered the comparisons related to the λ_α eigenvalue. The other non-negligible relative values are obtained for the $\lambda_{IR\beta0}$ element which curves are shown in Figure IV.16.

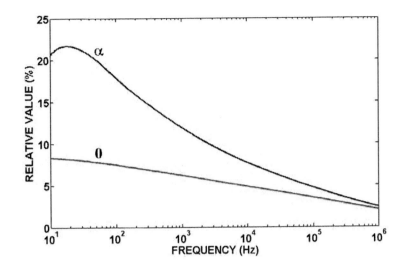

Figure IV.14. The $\lambda_{IR\alpha0}$ relative values for the actual asymmetrical vertical three-phase transmission line

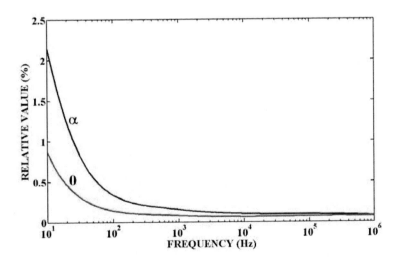

Figure IV.15. The $\lambda_{VR\alpha0}$ relative values for the actual asymmetrical vertical three-phase transmission line

In this case, the peak value is about 17 % and related to frequency values close to 20 Hz. It is in the initial of the considered frequency range. This frequency value is similar to that related to Figure IV.14.

Besides the peak of 17 % for frequency values close to 20 Hz, the range of relative values of the $\lambda_{IR\beta0}$ element is from 4 % to 18 %. This range is higher than that related to the $\lambda_{VR\beta0}$ element. For the $\lambda_{VR\beta0}$ element, the range of relative values is from 0.5 % to 4 %. Similar to the other both couplings, the T_I eigenvector change presents the highest relative value range in case of Figures IV.16 and IV.17.

The relative values in this item are calculated from the following:

$$\varepsilon_{IRKJ}(\%) = \frac{\lambda_{IRKJ}}{\lambda_K \ or \ \lambda_J} \cdot 100 \quad and \quad \varepsilon_{VKJ}(\%) = \frac{\lambda_{VRKJ}}{\lambda_K \ or \ \lambda_J} \cdot 100$$

(94)

$$K = \alpha, \beta, 0 \quad and \quad J = \alpha, \beta, 0 \quad and \quad K \neq J$$

Analyzing the results shown in the last six figures, the highest relative values are related to the elements that couple the 0 mode to the other both modes. So, considering these highest relative values associated to the $\lambda_{IR\alpha0}$ and $\lambda_{IR\beta0}$ elements, the relative value peaks of each element are associated to the mode different of the 0 mode. It is concluded that the influence of the 0 mode on the other both modes is more significant than the influence of the other both modes on the 0 mode. This is also observed for the typical symmetrical three-phase line investigated in the last item. If this is also observed for another typical asymmetrical three-phase transmission line, it is possible to carry out a correction procedure for Clarke's matrix based on the 0 mode.

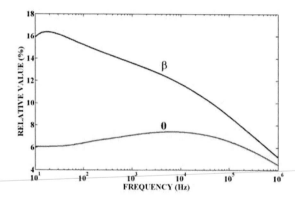

Figure IV.16. The $\lambda_{IR\beta0}$ relative values for the actual asymmetrical vertical three-phase transmission line

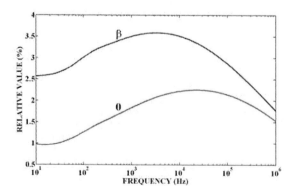

Figure IV.17. The $\lambda_{VR\beta0}$ relative values for the actual asymmetrical vertical three-phase transmission line

IV.4. Untransposed Three-Phase Lines with Phase Conductors Distributed in a Triangular Design

This line design, called asymmetrical triangular three-phase transmission line, is shown in the next figure. Because there are couplings among all quasi-modes, there are not null values in quasi-mode matrices as well as in longitudinal impedance and shunt admittance matrices in quasi-mode domain.

The relative errors are obtained using equations (86) and the next both figures show the results that depend on the frequency. Because the errors shown in the next both figures are considered negligible, the $\lambda_{I\Delta}$ and $\lambda_{V\Delta}$ quasi-mode matrices are described by equations (95) where the quasi-modes are changed into the correspondent λ eigenvalues.

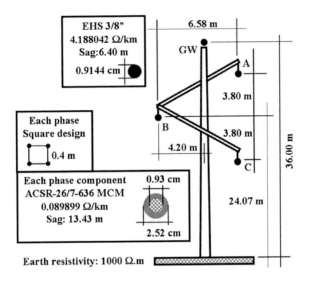

Figure IV.18. Actual asymmetrical triangular three-phase transmission line

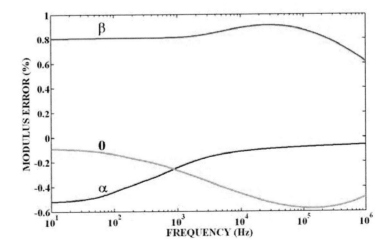

Figure IV.19. Comparisons among $\lambda_{I\Delta}$ quasi-modes and λ eigenvalues for the actual asymmetrical triangular three-phase transmission line

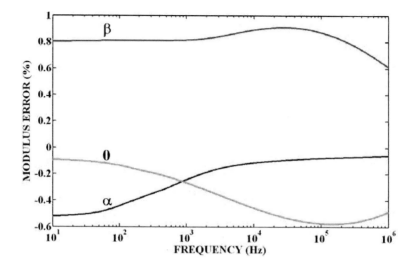

Figure IV.20. Comparisons among $\lambda_{V\Delta}$ quasi-modes and λ eigenvalues for the actual asymmetrical triangular three-phase transmission line

The $\lambda_{I\Delta}$ and $\lambda_{V\Delta}$ quasi-mode matrices are described by:

$$\lambda_{I\Delta} = \begin{bmatrix} \lambda_{\alpha} & \lambda_{I\Delta\alpha\beta} & \lambda_{I\Delta\alpha0} \\ \lambda_{I\Delta\alpha\beta} & \lambda_{\beta} & \lambda_{I\Delta\beta0} \\ \lambda_{I\Delta\alpha0} & \lambda_{I\Delta\beta0} & \lambda_0 \end{bmatrix} \quad and \quad \lambda_{V\Delta} = \begin{bmatrix} \lambda_{\alpha} & \lambda_{V\Delta\alpha\beta} & \lambda_{V\Delta\alpha0} \\ \lambda_{V\Delta\alpha\beta} & \lambda_{\beta} & \lambda_{V\Delta\beta0} \\ \lambda_{V\Delta\alpha0} & \lambda_{V\Delta\beta0} & \lambda_0 \end{bmatrix} \quad (95)$$

The impedance and admittance errors are the next both figures.

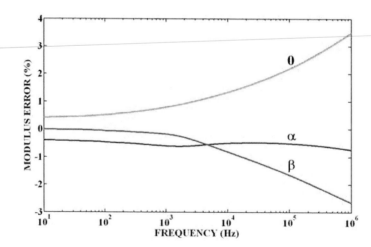

Figure IV.21. Relative errors related to the mode longitudinal impedances for the actual asymmetrical triangular three-phase transmission line

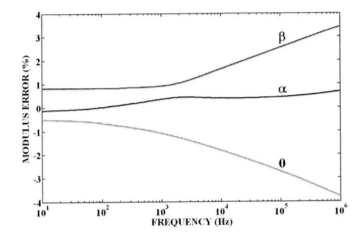

Figure IV.22. Relative errors related to the mode shunt admittances for the actual asymmetrical triangular three-phase transmission line

The longitudinal impedance and shunt admittance matrices in quasi-mode domain are described by next equations and based on the relative errors shown in the next both figures. The proposed consideration of the quasi-mode values as the mode ones depends on the application and the considered frequency, because the error absolute values in the last both figures are increasing for increasing frequency values.

$$
Z_{M\Delta} = \begin{bmatrix} Z_\alpha & Z_{\alpha\beta} & Z_{\alpha 0} \\ Z_{\alpha\beta} & Z_\beta & Z_{\beta 0} \\ Z_{\alpha 0} & Z_{\beta 0} & Z_0 \end{bmatrix} \quad and \quad Y_{M\Delta} = \begin{bmatrix} Y_\alpha & Y_{\alpha\beta} & Y_{\alpha 0} \\ Y_{\alpha\beta} & Y_\beta & Y_{\beta 0} \\ Y_{\alpha 0} & Y_{\beta 0} & Y_0 \end{bmatrix} \tag{96}
$$

In case of off-diagonal elements of the $\lambda_{I\Delta}$ and $\lambda_{V\Delta}$ quasi-mode matrices, the relative values are calculated using equations (97) and the results are shown in the next six figures.

$$
\varepsilon_{I\Delta KJ}(\%) = \frac{\lambda_{I\Delta KJ}}{\lambda_K \ or \ \lambda_J} \cdot 100 \quad and \quad \varepsilon_{VKJ}(\%) = \frac{\lambda_{V\Delta KJ}}{\lambda_K \ or \ \lambda_J} \cdot 100 \tag{97}
$$

$$
K = \alpha, \beta, 0 \quad and \quad J = \alpha, \beta, 0 \quad and \quad K \neq J
$$

The comparisons of $\lambda_{I\Delta\alpha\beta}$ and $\lambda_{V\Delta\alpha\beta}$ quasi-mode couplings to the correspondent eigenvalues are shown in the next both figures, respectively. The relative values related to the T_I eigenvector change into Clarke's matrix ($\lambda_{I\Delta\alpha\beta}$) are higher than those related to the T_V eigenvector change ($\lambda_{V\Delta\alpha\beta}$). The relative value peak in these both figures is about 0.27 % and related to a frequency value about 20 kHz. The relative values shown in these figures can be considered negligible.

Analyzing the $\lambda_{I\Delta\alpha 0}$ and $\lambda_{V\Delta\alpha 0}$ relative values, in Figures IV.25 and IV.26, it is shown results that can also be considered negligible. The relative value peak in these mentioned figures is about 2.4 % and it is related a frequency value close to 20 Hz. The quasi-mode coupling relative values based on the YZ product, which is related to the T_I eigenvector, are higher than those based on the ZY matricial product, which is related to the T_V eigenvector. This characteristic is similar to the other analyzed relative values in this item and in the previous both items.

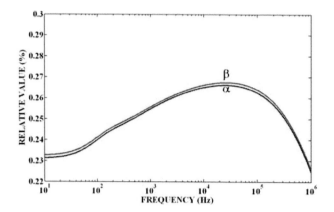

Figure IV.23. The $\lambda_{I\Delta\alpha\beta}$ relative values for the actual asymmetrical triangular three-phase transmission line

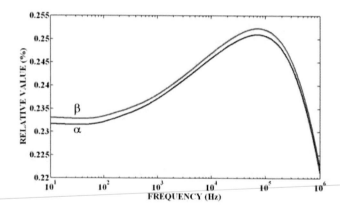

Figure IV.24. The $\lambda_{V\Delta\alpha\beta}$ relative values for the actual asymmetrical triangular three-phase transmission line

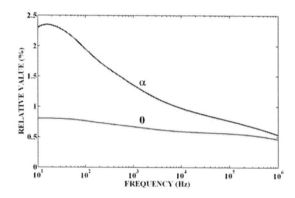

Figure IV.25. The $\lambda_{I\Delta\alpha0}$ relative values for the actual asymmetrical triangular three-phase transmission line

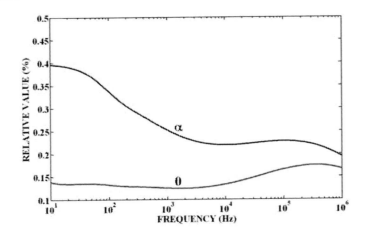

Figure IV.26. The $\lambda_{V\Delta\alpha0}$ relative values for the actual asymmetrical triangular three-phase transmission line

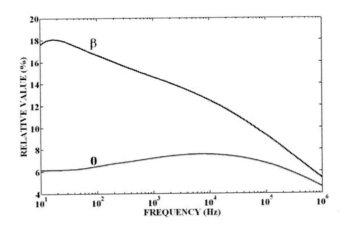

Figure IV.27. The $\lambda_{I\Delta\beta0}$ relative values for the actual asymmetrical triangular three-phase transmission line

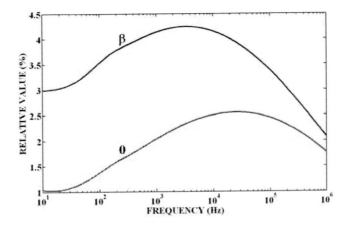

Figure IV.28. The $\lambda_{V\Delta\beta0}$ relative values for the actual asymmetrical triangular three-phase transmission line

For the asymmetrical triangular three-phase transmission line shown in this item, the highest relative values of off-diagonal elements of the $\lambda_{I\Delta}$ and $\lambda_{V\Delta}$ matrices are the $\lambda_{I\Delta\beta0}$ ones. These values are shown in Figure IV.27 and the peak value is 18 % approximately. This peak is in the initial of the considered frequency range. The Figure IV.28 shows the $\lambda_{V\Delta\beta0}$ relative values, completing the analyses of this item.

The $\lambda_{V\Delta\beta0}$ relative values are lower than the $\lambda_{I\Delta\beta0}$ ones. This is a characteristic property for all relative value couplings among quasi-modes for the analyzed three typical three-phase transmission lines. So, the off-diagonal elements of the quasi-mode matrices obtained from the YZ matricial product have higher relative values than those obtained from ZY matricial product. For untransposed symmetrical and asymmetrical three-phase transmission lines, changing the T_I eigenvector into Clarke's matrix, no negligible couplings among the quasi-modes can be obtained and it can lead to significant errors, if the quasi-mode values are used to calculated other variables in mode domain, neglecting the mentioned couplings. On the other hand, the main advantage associated to the mode domain is the manipulation of diagonal matrices where only the elements of matrix main diagonal are not null. This implies that it is necessary to obtain a better approximation for the T_I eigenvector. An alternative is to apply a correction procedure to Clarke's matrix, improving the approximation for the T_I eigenvector. It is carried out in the next item.

V. CORRECTION PROCEDURE APPLICATION TO CLARKE'S MATRIX [42]

The correction procedure for Clarke's matrix is analyzed step by step in this section, searching for details of each matrix that composes this procedure and considering the typical three-phase transmission lines analyzed previously. Changing the eigenvectors as modal transformation matrices, Clarke's matrix has chosen because this matrix is an eigenvector matrix for transposed symmetrical and asymmetrical three-phase transmission lines. In this chapter, the application of this matrix has been analyzed considering untransposed symmetrical and asymmetrical three-phase transmission lines. In most of these cases, the errors related to the eigenvalues can be considered negligible. It is not true when it is analyzed the elements that are not in main diagonal of the matrix obtained from the Clarke's matrix application. The obtained matrix is correspondent to the eigenvalue matrix and called quasi-mode matrix. Its off-diagonal elements represent couplings among the quasi-modes. So, the off-diagonal quasi-mode element relative values are not negligible when compared to the eigenvalues that correspond to the coupled quasi-modes. Minimizing these relative values, the correction procedure is analyzed in detail.

In mode domain, the frequency influence can be easily introduced on the line parameters. Using phase-mode transformation matrices, all electrical parameters and all line representative matrices are obtained in mode domain [1, 2, 3, 4]. The line representative matrices become diagonal and the frequency influence can independently be introduced for every mode. Applying frequency dependent line parameters also leads to frequency dependent phase-mode transformation matrices. For obtaining voltages and currents in phase domain, it is necessary to use convolution procedures [5, 6, 7, 8, 9, 10 14]. An alternative is to change the exact transformation matrices into single real ones and, then, any values can be determined in phase or mode domain using only a matricial multiplication [3, 11-14].

Modal transformations have been applied to three phase transmission lines using Clarke's matrix based on fact that this matrix is an eigenvector one for transposed symmetrical and asymmetrical three-phase transmission lines. For unstransposed three-phase transmission line cases, when eigenvector matrices are changing into Clarke's matrix, the resulting matrix correspondent to the eigenvalue one is not a diagonal matrix. The result of the Clarke's matrix application are called quasi-mode matrix and the elements of this matrix that are not in the main diagonal represent couplings among the quasi-modes. The off-diagonal quasi-mode relative values are not negligible, if they are compared to the exact modes correspondent to the coupled quasi-modes.

Based on these results, the Clarke's matrix application is analyzed considering symmetrical three-phase lines and a frequency range from 10 Hz to 1 MHz. The quasi-mode errors related to the eigenvalues are studied as well as the off diagonal elements of the quasi-mode matrix. Searching for the off-diagonal element relative value minimization, a perturbation approach corrector matrix is applied to Clarke's matrix. This correction procedure application is analyzed step by step considering the errors related to the eigenvalues and the off-diagonal element relative values.

V.1. The Perturbation Approach Corrector Matrix [5]

The procedure shown in this section is based on a first order perturbation theory approach [3]. This procedure improves the quasi-mode results and obtains a better approximation to the exact values. Initializing, the λ_{IS}, λ_{VS}, λ_{IR}, λ_{VR}, λ_{IA}, λ_{VA} matrices are portioned into two blocks. In this section, these matrices are represented by the λ_P matrix.

$$\lambda_P = \begin{bmatrix} & \lambda_{P22} & \begin{matrix} \lambda_{P\alpha 0} \\ \lambda_{P\beta 0} \end{matrix} \\ \lambda_{P0\alpha} & \lambda_{P0\beta} & \lambda_{P0} \end{bmatrix} \tag{98}$$

The λ_{P22} matrix structure is:

$$\lambda_{P22} = \begin{bmatrix} \lambda_{P\alpha} & \lambda_{P\alpha\beta} \\ \lambda_{P\beta\alpha} & \lambda_{P\beta} \end{bmatrix} \tag{99}$$

Although the line representative matrices are symmetrical, small numeric differences are considered among symmetrical elements of the λ_P matrix. Based on the last two equations, the A_α and A_β elements are determined by:

$$A_\alpha = \frac{\text{trace}(\lambda_{P22}) + \sqrt{\text{trace}^2(\lambda_{P22}) - 4 \cdot \det(\lambda_{P22})}}{2}$$

$$A_\beta = \frac{\text{trace}(\lambda_{P22}) - \sqrt{\text{trace}^2(\lambda_{P22}) - 4 \cdot \det(\lambda_{P22})}}{2}$$

(100)

The trace term in the last equation is related to the sum of the diagonal elements of the considered matrix. The det term is the determinant of the considered matrix.

The n_{21} and n_{12} values are determined by:

$$n_{21} = \frac{A_\alpha - \lambda_{P\alpha}}{\lambda_{P\alpha\beta}} \quad and \quad n_{12} = \frac{A_\beta - \lambda_{P\beta}}{\lambda_{P\beta\alpha}}$$

(101)

Applying these elements, it is obtained the N_{22} matrix.

$$N_{22} = \begin{bmatrix} 1 & n_{12} \\ n_{21} & 1 \end{bmatrix}$$

(102)

Using the N_{22} matrix, it is determined a normalization matrix:

$$N = \begin{bmatrix} N_{22} & 0 \\ 0 & 1 \end{bmatrix}$$

(103)

In the next equation, the N matrix is applied to the λ_{VS}, λ_{VR}, λ_{VA} matrices matrix for calculating of the A matrix considering the correction applied to the T_V matrix.

$$A = N^{-1} \cdot T_{CL}^T \cdot Z \cdot Y \cdot T_{CL} \cdot N = \begin{bmatrix} A_\alpha & 0 & A_{\alpha 0} \\ 0 & A_\beta & A_{\beta 0} \\ A_{0\alpha} & A_{0\beta} & A_0 \end{bmatrix}$$

(104)

Considering λ_{IS}, λ_{IR}, λ_{IA} matrices and applying the correction procedure for the T_I matrix, the A matrix determination is similar to the T_V one, changing the position of the Z and Y matrices.

$$A = N^{-1} \cdot T_{CL}^T \cdot Y \cdot Z \cdot T_{CL} \cdot N = \begin{bmatrix} A_\alpha & 0 & A_{\alpha 0} \\ 0 & A_\beta & A_{\beta 0} \\ A_{0\alpha} & A_{0\beta} & A_0 \end{bmatrix}$$

(105)

The structure of the A matrix is determined from:

$$\lambda = A + \left(\lambda_{CL} \cdot Q - Q \cdot \lambda_{CL}\right) \qquad (106)$$

The λ_{CL} is the eigenvalue matrix calculated using Clarke's matrix as an eigenvector matrix and considering the analyzed three-phase transmission line in ideally transposition. This can be obtained with only one coupling value in each Z and Y matrices as well as only one value self value in the mentioned matrices. These unique values can be obtained from the arithmetic media of the correspondent values of the matrices related to the untransposed situation of the considered transmission line.

The last equation leads to:

$$\begin{cases} \lambda_K = A_K, & K = \alpha, \beta, 0 \\ A_{JK} = \left(\lambda_{CL\,K} - \lambda_{CL\,J}\right) \cdot Q_{JK}, & J \neq K \end{cases} \qquad (107)$$

The A_α and A_β elements have already calculated in this section. The last equation is used for calculating the A_0 element, as well as, the Q_{JK} elements. Because the $\lambda_{CL\,\alpha}$ element is equal to the $\lambda_{CL\,\beta}$, this implies that the $A_{\alpha\beta}$, $A_{\beta\alpha}$, $Q_{\alpha\beta}$ and $Q_{\beta\alpha}$ elements are null. Then, only the Q matrix elements of the third line and the third row are not null. These elements correspond to the 0 mode and are calculated by:

$$Q_{0K} = \frac{A_{0K}}{\lambda_{CL\,K} - \lambda_{CL\,0}} \quad and \quad Q_{K0} = \frac{A_{K0}}{\lambda_{CL\,0} - \lambda_{CL\,K}} \qquad (108)$$

$$K = \alpha, \beta$$

The Q matrix structure is:

$$Q = \begin{bmatrix} 0 & 0 & Q_{\alpha 0} \\ 0 & 0 & Q_{\beta 0} \\ Q_{0\alpha} & Q_{0\beta} & 0 \end{bmatrix} \qquad (109)$$

The perturbation approach corrector matrix is described by:

$$W = N \cdot \left(I + Q\right) \quad and \quad W^{-1} = \left(I + Q^{-1}\right) \cdot N^{-1} \qquad (110)$$

The corrected transformation matrices are described by:

$$T_{NV} = W^{-1} \cdot T_{CL}^{T} \quad and \quad T_{NV}^{-1} = T_{CL} \cdot W \qquad (111)$$

V.2. Flowchart of the Correction Procedure

Analyzing step by step the correction procedure, each matrix that composed the W one is applied and its influence is shown. It is studied the contribution of each W term for minimization of errors and the relative values of the off-diagonal quasi-mode matrix elements. Figure V.1 presents the flowchart for these analyses. Checking the changes into the Clarke's matrix, results carried out for the λ_{NCL} matrix representing an untransposed symmetrical three-phase transmission line [15-17].

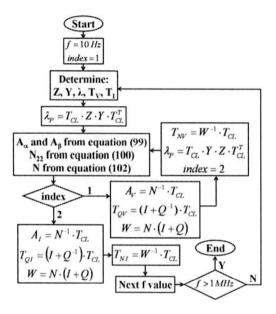

Figure V.1. Correction procedure flowchart

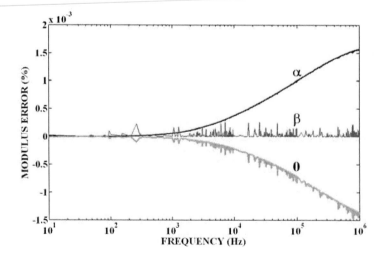

Figure V.2. The quasi-mode relative errors obtained with the Q matrix application for the actual symmetrical three-phase transmission line.

V.3. The Q Matrix Application

Checking the changes into the Clarke's matrix results carried out applying only the Q matrix, it is not used the N and W matrices for corrections based on the T_I and T_V eigenvectors shown in the flowchart of Figure V.1. In this case, only the Q matrix is applied as a correction matrix to Clarke's matrix for checking this matrix influence on the proposed correction. Figures V.2, V.3 and V.4 are associated to the Q matrix application to the quasi-modes related to the untransposed symmetrical three-phase line shown in Figure III.4. So, Figure V.2 is associated to the relative errors of the quasi-modes. Figure V.3 shows the relative values of the $\lambda_{IS\alpha0}$ coupling and Figure V.4 is related to the $\lambda_{VS\alpha0}$ relative values. In case of relative errors, the results related to the T_V eigenvector are equal to those related to the T_I eigenvector. Because of this, only one set of curves is shown.

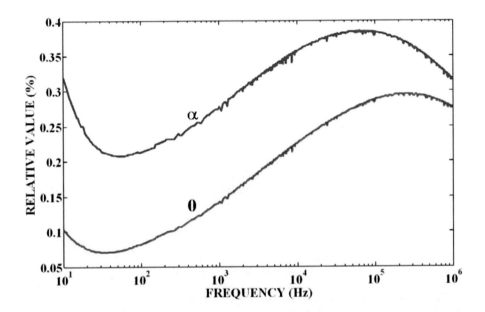

Figure V.3. The $\lambda_{IS\alpha0}$ relative values after the Q matrix application

Analyzing only the results related to the symmetrical three-phase transmission line, it is concluded that the Q reduces the quasi-mode relative errors expressively. The decrease is about 150 times when the results obtained for the Q matrix application are compared to those of Figures IV.1 and IV.2. In case of the $\lambda_{IS\alpha0}$ relative values, the decrease of peak value is about 45 times comparing to the peak value of the Figure IV.5. For the $\lambda_{VS\alpha0}$ relative values, it is obtained a 17 time reduction for peak values, approximately, when it is compared the results of Figure V.4 to those related to Clarke's matrix shown in Figure IV.6.

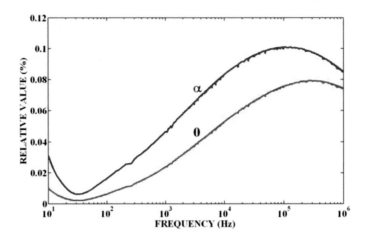

Figure V.4. The $\lambda_{VS\alpha 0}$ relative values after the Q matrix application

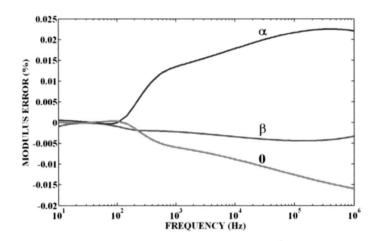

Figure V.5. The quasi-mode relative errors obtained with the Q matrix application for the actual asymmetrical vertical three-phase transmission line

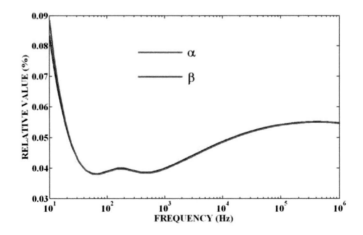

Figure V.6. The $\lambda_{IR\alpha\beta}$ relative values after the Q matrix application

For asymmetrical vertical three-phase transmission line that is shown in Figure IV.7, after the Q matrix application to correction of Clarke's matrix, the quasi-mode relative error range is decreased about 30 times comparing to Figures IV.8 and IV.9. Besides this, the relative error curve shapes are modified and the β component has values closer to null value than the other components. Previously, this characteristic is associated to the α quasi-mode.

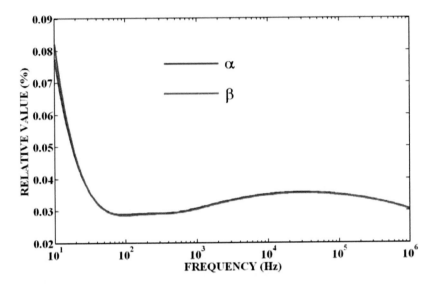

Figure V.7. The $\lambda_{VR\alpha\beta}$ relative values after the Q matrix application.

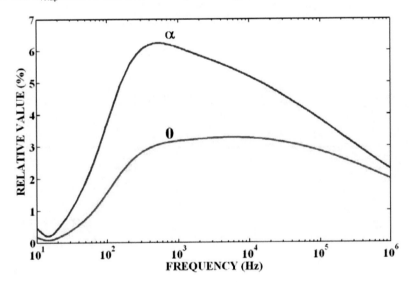

Figure V.8. The $\lambda_{IR\alpha0}$ relative values after the Q matrix application.

Applying the Q matrix and considering the $\lambda_{IR\alpha\beta}$ relative values, it is observed a 5 time decrease in the peak value when it is compared Figures V.6 and IV.12. Besides this, the curve shapes are changed. For the $\lambda_{VR\alpha\beta}$ relative values, the curve shapes are not changed expressively and the reduction of the peak value is about 2 times when it is compared Figure V.7 to Figure IV.13.

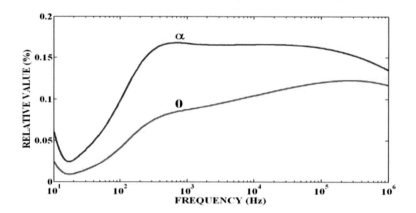

Figure V.9. The $\lambda_{VR\alpha0}$ relative values after the Q matrix application.

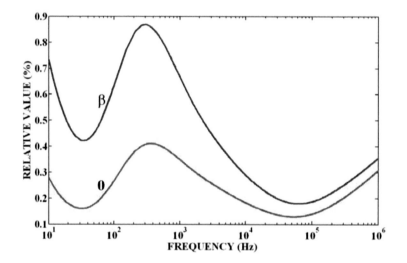

Figure V.10. The $\lambda_{IR\beta0}$ relative values after the Q matrix application.

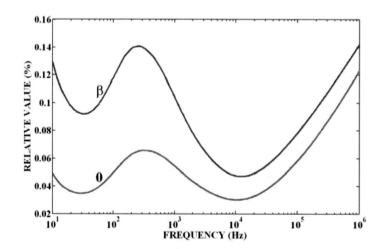

Figure V.11. The $\lambda_{VR\beta0}$ relative values after the Q matrix application

Analyzing the $\lambda_{IR\alpha0}$ relative values after the Q matrix application in Figure V.8, it is observed that these relative values are decreased about 3.5 times and the curve shapes are modified when compared those curves shown in Figure IV.14. For the $\lambda_{VR\alpha0}$ relative values shown in Figure V.9 when they are compared to those values shown in Figure IV.15, it is noted modifications in the curve shapes and the there is a 10 time reduction in the peak value, approximately.

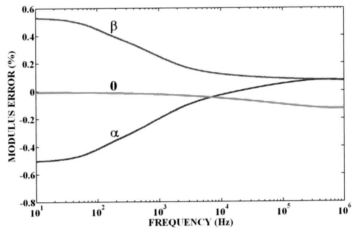

Figure V.12 – The quasi-mode relative errors obtained with the Q matrix application for the actual asymmetrical triangular three-phase transmission line

Considering the $\lambda_{IR\beta0}$ and $\lambda_{VR\beta0}$ relative values obtained from the Q matrix application, there are changes in the shapes of the curves comparing Figures V.10 and V.11 to the Figures IV.16 and IV.17, respectively. For the $\lambda_{IR\beta0}$ relative values, the peak value is reduced from about 17 % to 0.9 %. It is a 18 time reduction, approximately. Considering the $\lambda_{VR\beta0}$ relative values, the reduction is about 25 times.

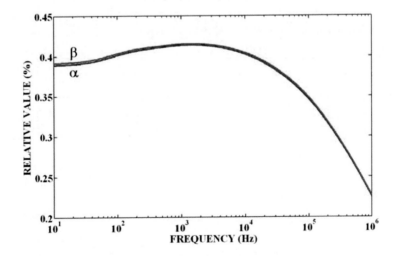

Figure V.13. The $\lambda_{1A\alpha\beta}$ relative values after the Q matrix application.

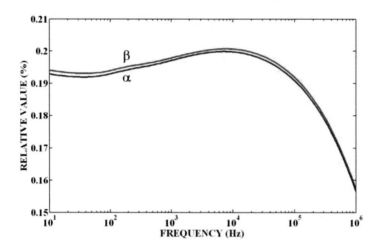

Figure V.14. The $\lambda_{V\Delta\alpha\beta}$ relative values after the Q matrix application.

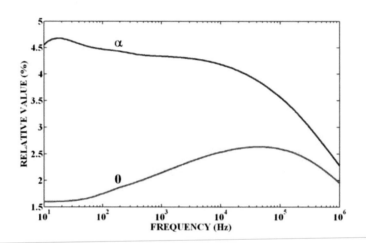

Figure V.15. The $\lambda_{I\Delta\alpha 0}$ relative values after the Q matrix application.

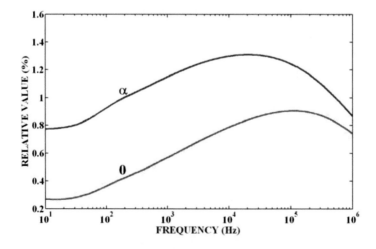

Figure V.16. The $\lambda_{V\Delta\alpha 0}$ relative values after the Q matrix application

For the asymmetrical triangular three-phase transmission line, in Figure V.12, the 0 quasi-mode curve is closer to null value than the same component before the Q matrix application (Figures IV.19 and IV.20). For the other quasi-mode curves, they tend to decrease with the frequency increasing. This characteristic is not observed previously. The modulus error range is not expressively decreased.

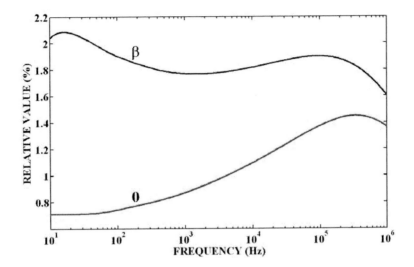

Figure V.17. The $\lambda_{I\Delta\beta0}$ relative values after the Q matrix application

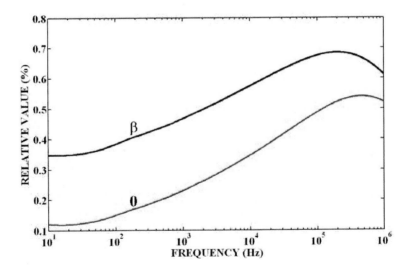

Figure V.18. The $\lambda_{V\Delta\beta0}$ relative values after the Q matrix application.

In Figure V.13, it is shown the $\lambda_{I\Delta\alpha\beta}$ relative values after the Q matrix application. The relative value peak decreases about 1.3 times. Because, in Figure IV.23, these values have already been small, the changes in the values are not significant.

Similar conclusions can be associated to the comparisons between Figure V.14 and Figure IV.24, because the decrease is about 1.2 times for peak value.

For the $\lambda_{I\Delta\alpha0}$ and $\lambda_{V\Delta\alpha0}$ relative values after the Q matrix application, there are changes in the curves shapes, considering the comparisons of Figures V.15 and V.16 to Figures IV.25 and IV.26. While for other couplings related to this analyzed line as well as to the other both lines used in this chapter, for $\lambda_{I\Delta\alpha0}$ and $\lambda_{V\Delta\alpha0}$ relative values, the peak values are increased.

Comparing Figures V.15 and IV.25, it is verified a 1.9 time increasing, approximately, while for $\lambda_{V\Delta\alpha0}$ relative values, this increasing is about 3.2 times comparing Figures V.16 and IV.26.

Considering the $\lambda_{I\Delta\beta0}$ and $\lambda_{I\Delta\beta0}$ relative values after the Q matrix application, there are also changes in the curve shapes and the value peaks are also decreased. For $\lambda_{I\Delta\beta0}$ relative values, it is decreased about from 18 % to 2.1 %. It is about a 8 time decreasing. In this case, it is compared Figures V.17 and IV.27.

For $\lambda_{V\Delta\beta0}$ values, it is decreased about 6 times, from 4 % to 0.7 % and it is considering Figures V.18 and IV.28.

The Q matrix application is capable to decrease the all quasi-mode relative modulus errors obtained from the application of Clarke's matrix for all three typical three-phase transmission lines analyzed in this paper. In this case, these lines are considered untransposed. For the off-diagonal element relative values of quasi-mode matrices, the most of these values are reduced with the Q matrix application. Only for the $\lambda_{I\Delta\alpha0}$ and $\lambda_{V\Delta\alpha0}$ values, that are related to the untransposed asymmetrical triangular three-phase transmission line, it is verified that these relative values are increased after the Q matrix application. Based on the shown equations, for completing the correction procedure, it is applied the N matrix. It is analyzed in the next item.

V.4. The N Matrix Application for Completing the Correction Procedure

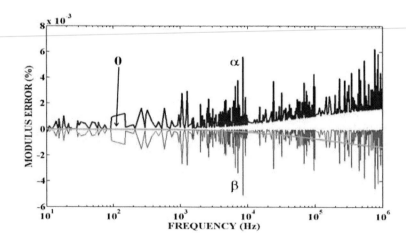

Figure V.19. The λ_{IS} quasi-mode errors after the correction procedure application for the actual symmetrical three-phase transmission line

After the Q matrix use, the correction procedure is completed applying the N matrix. The main function of this matrix is to decrease some coupling values that has had no negligible relative values yet. Considering the symmetrical three-phase transmission line, the relative modulus errors present some differences when they are calculated based on the T_I eigenvectors and the T_V ones. Figure V.19 shows the results associated to the T_I eigenvectors and Figure V.20 shows the results related to the T_V eigenvectors. Comparing the both next figures, there are differences in the curve shapes and in the superior border of the error range. In both figures, the α and β curves present numeric oscillations. For Figure V.19, the decrease of the relative errors is about 35 times when compared to Figure IV.1. Considering Figure V.20, for superior border of vertical axis, the reduction is about 50 times and, for inferior border of this axis, it is about 33 times.

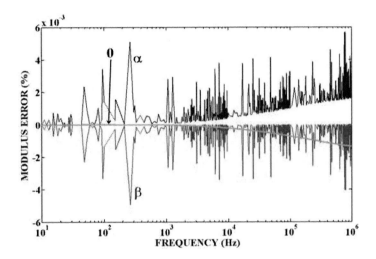

Figure V.20. The λ_{VS} quasi-mode errors after the correction procedure application for the actual symmetrical three-phase transmission line

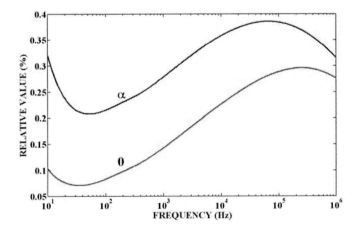

Figure V.21. The $\lambda_{IS\alpha0}$ relative values after the correction procedure application

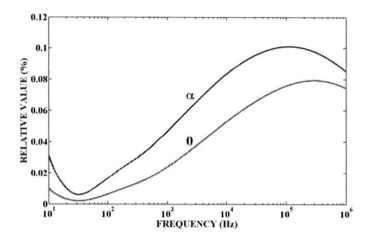

Figure V.22. The $\lambda_{VS\alpha 0}$ relative values after the correction procedure application

Considering the single coupling for the symmetrical three-phase transmission line and comparing Figures V.21 and IV.5, the $\lambda_{IS\alpha 0}$ relative values have a 45 time reduction, approximately. For $\lambda_{VS\alpha 0}$ relative values, it is decreased about 17 times when compared Figures V.22 and IV.6. For both figures, V.21 and V.22, the curve shapes are different of those obtained with the application of Clarke's matrix. The frequency values of the peak values are also different, comparing to the results obtained from Clarke's matrix.

Figure V.23 is related to the asymmetrical vertical three-phase transmission line. Comparing this figure to Figure IV.8, it is observed that there are changes in the curve shapes. The reduction of the relative modulus error range is about 25 times. In this case, there are not differences between the results associated to the T_I eigenvectors and those associated to the T_V eigenvectors.

The $\alpha\beta$ coupling results are shown in Figures V.24 and V.25.

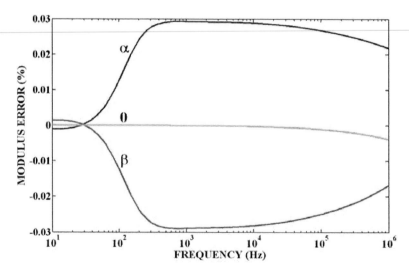

Figure V.23. The relative modulus errors after the correction procedure application for the actual asymmetrical vertical three-phase transmission line

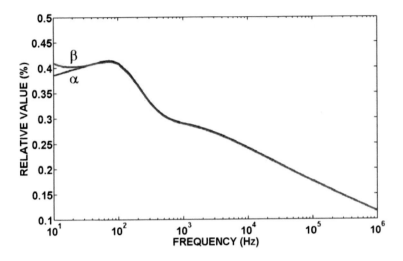

Figure V.24. The $\lambda_{IR\alpha\beta}$ relative values after the correction procedure application

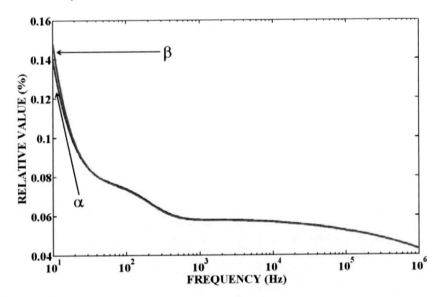

Figure V.25. The $\lambda_{VR\alpha\beta}$ relative values after the correction procedure application

In Figure V.24, it is observed significant changes only in the curve shapes when compared to Figure IV.12. There is not significant reduction of the relative values. On the other hand, comparing Figures V.25 and IV.13, the changes are not significant for the values and the curve shapes.

The next both figures are related to the $\lambda_{IR\alpha 0}$ and $\lambda_{VR\alpha 0}$ couplings and they are compared to Figures IV.14 and IV.15, respectively. Only for $\lambda_{IR\alpha 0}$ relative values, it is observed significant changes in the curve shapes.

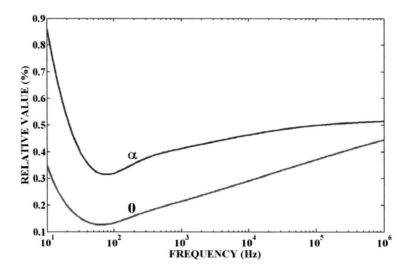

Figure V.26. The $\lambda_{IR\alpha 0}$ relative values after the correction procedure application

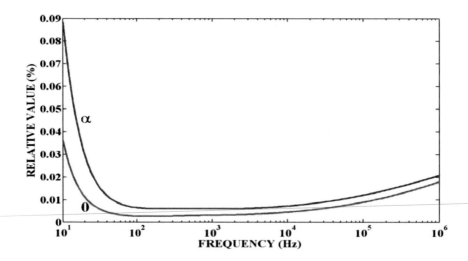

Figure V.27. The $\lambda_{VR\alpha 0}$ relative values after the correction procedure application.

Considering the comparisons associated to the $\lambda_{IR\alpha 0}$ curves, the reduction is about 25 times, while, for the $\lambda_{VR\alpha 0}$ curves, it is about 22 times.

The reduction for the asymmetrical vertical three-phase transmission line, when compared the results obtained from Clarke's matrix and those obtained after the correction procedure application, can be associated to a characteristic value of 25 times. So, comparing Figures V.28 and IV.16, the reduction is about 25 times. In this case, there are also changes in the curve shapes.

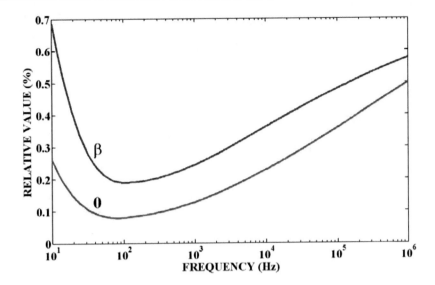

Figure V.28. The $\lambda_{IR\beta 0}$ relative values after the correction procedure application

For the $\lambda_{VR\beta 0}$ relative values, comparing Figures V.29 and IV.17, the relative values are decreased about 17 times and there are changes in the curve shapes.

Considering the asymmetrical triangular three-phase transmission line, Figure V.30 shows the relative modulus error curves that is compared to Figure IV.19. With the correction procedure, it is obtained a 2 time reduction for the superior border of the vertical axis, approximately. For the inferior border, the reduction is about 1.25 times.

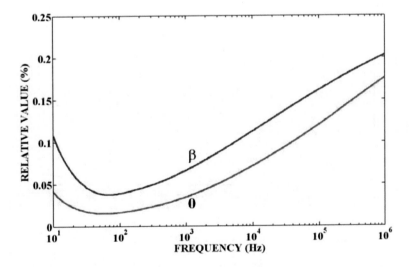

Figure V.29. The $\lambda_{VR\beta 0}$ relative values after the correction procedure application

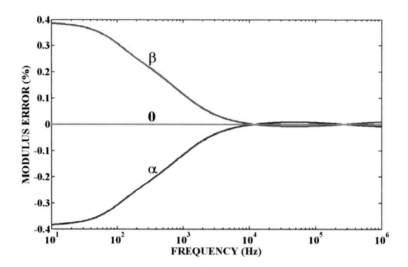

Figure V.30. The relative modulus errors after the correction procedure application for the actual asymmetrical vertical three-phase transmission line

There are significant changes in curve shapes and in the relative values for high frequencies, when it is compared Figures V.31 and IV.23 as well as Figures V.32 and IV.24. For high frequencies, the reductions of relative values reach about 4.5 times.

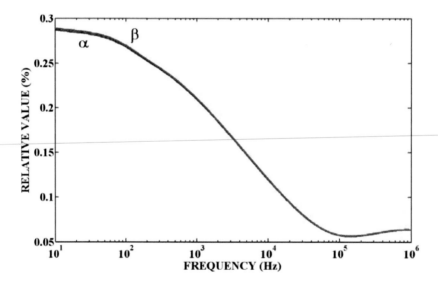

Figure V.31. The $\lambda_{1\Delta\alpha\beta}$ relative values after the correction procedure application

A 110 time reduction on the peak of the relative values and changes in the curve shapes can be observed comparing Figures V.33 and IV.25. For the $\lambda_{V\Delta\alpha0}$ relative values, it is observed changes in the curve shapes and the reduction reaches about 50 times.

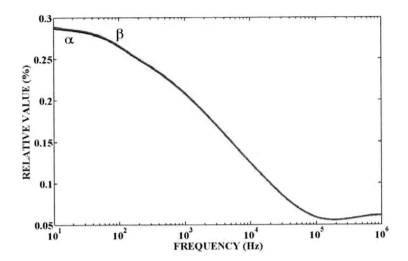

Figure V.32. The $\lambda_{V\Delta\alpha\beta}$ relative values after the correction procedure application

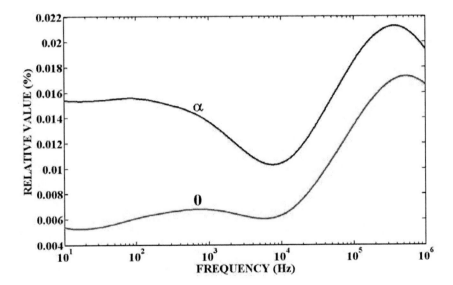

Figure V.33. The $\lambda_{I\Delta\alpha0}$ relative values after the correction procedure application

For all three typical transmission line samples shown in this chapter, the highest absolute reduction is obtained for the $\lambda_{I\Delta\beta0}$ relative values with the correction procedure application. This reduction is about 140 times. Besides this, there are changes in the curve shapes when Figs V.35 and IV.25 are compared.

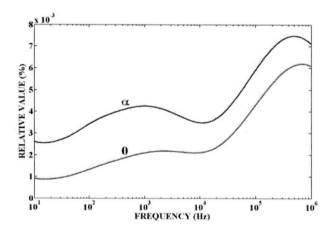

Figure V.34. The $\lambda_{V\Delta\alpha0}$ relative values after the correction procedure application

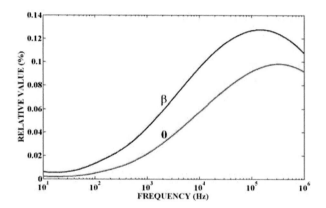

Figure V.35. The $\lambda_{I\Delta\beta0}$ relative values after the correction procedure application

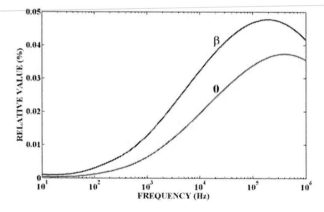

Figure V.36. The $\lambda_{V\Delta\beta0}$ relative values after the correction procedure application

Finally, considering Figures V.36 and IV.26, it is shown the last comparisons of this item. In this case, the reduction reaches about 90 times and the curve shapes are also modified.

For general analyses, the correction procedure application to Clarke's matrix reduces the relative quasi-mode modulus errors and the off-diagonal relative values of the quasi-mode couplings. With this procedure, it is also obtained balanced relative modulus errors and balanced off-diagonal relative values.

Because the correction procedure is based on Clarke's matrix that is composed by real and constant elements, the both transformations matrices obtained for each analyzed case, one for voltage transformations and the other for the current transformations, are composed by complex elements which imaginary parts are small when compared to their absolute values and real parts. Besides of this, these elements are smoothly influenced by the frequency. These characteristics can be used for creating interesting applications for electromagnetic transient simulations. If the interest is the application of the exact transformation matrices, the obtained matrices from the correction procedure can be used, because the all relative errors and relative values can be considered negligible. Probably, this will be associated to convolution numeric methods. Using the both corrected transformation matrices for a specific frequency, it avoids the convolution numeric methods and frequency scan analyses could be used for validating this application. Simplifying much more, only the real part of the obtained both corrected transformations matrices can be considered. For frequency scan analyses, this alternative has presented negligible differences when compared to the application of the both corrected transformation matrices, considering a symmetrical three-phase transmission line. Based on the obtained results, because the new both transformations matrices can be considered exact transformation matrices, having small imaginary parts and smooth influence of the frequency, it is possible to get some other simplifications from them.

VI. IMPROVING SPECIFIC RESULTS [39-41]

If the analyzed three-phase transmission line is symmetry or, at least, if it has a phase conductor in the horizontal center of the phase conductors, it is possible to use a modified version of equations (110) and (111), improving some results related to the relative errors and the off-diagonal relative values. In this case, based on the T_V eigenvectors, it is applied the following equations:

$$W_S = N \cdot (I + Q) \cdot N^{-1} \quad and \quad W_S^{-1} = N \cdot (I + Q^{-1}) \cdot N^{-1} \qquad (112)$$

Now, the corrected transformation matrices are described by:

$$T_{NVS} = W_S^{-1} \cdot T_{CL}^T \quad and \quad T_{NVS}^{-1} = T_{CL} \cdot W_S \qquad (113)$$

Basing on the T_I eigenvectors, it is necessary to change the positions of the Z and Y matrices according to demonstrate in the item V.1 of this chapter. The improved results are the relative modulus errors associated to the symmetrical and asymmetrical three-phase transmission lines. For the asymmetrical vertical three-phase transmission line shown in this chapter, the application of the modified mentioned equations cause more reductions of the

$\lambda_{IR\alpha\beta}$ and $\lambda_{VR\alpha\beta}$ relative values without affecting the other ones when compared to the application of the equations (110) and (111).

The next both figures show the improved relative modulus errors related to the symmetrical and asymmetrical vertical three-phase transmission lines, respectively.

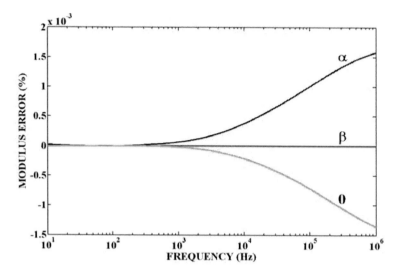

Figure VI.1. The relative modulus errors obtained with the specific correction procedure for the actual symmetrical three-phase transmission line

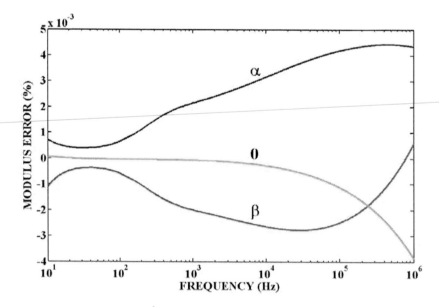

Figure VI.2. The relative modulus errors obtained with the specific correction procedure for the actual asymmetrical vertical three-phase transmission line

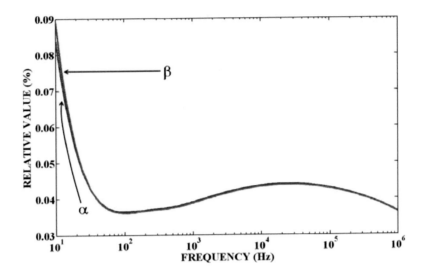

Figure VI.3. The improved $\lambda_{IR\alpha\beta}$ relative values

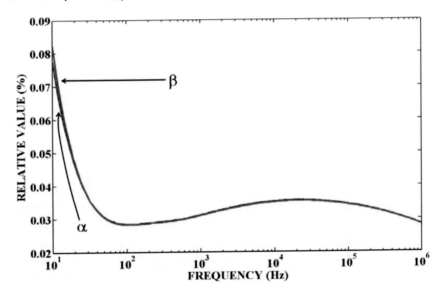

Figure VI.4. The improved $\lambda_{VR\alpha\beta}$ relative values

Completing the analyses of this item, it is shown in Figure VI.3 and VI.4, the improved $\lambda_{IR\alpha\beta}$ and $\lambda_{VR\alpha\beta}$ relative values, respectively.

Comparing Figure VI.1 to Figures V.19 and V.20, the main changes are that the numeric oscillations are removed. Considering the Figure VI.2, it is obtained a 6 time additional reduction for the peak value when it is compared to Figure V.23. Considering the $\lambda_{IR\alpha\beta}$ and $\lambda_{VR\alpha\beta}$ relative values, the additional reduction reaches about 5 times and 1.75 times for peak values, respectively.

VI.1. Other Analyses Associated to the Applications Shown in this Chapter

It has been investigated the inclusion of the technique that reduces the complex elements of the modal transformation matrices in a numeric routine based on state variables. In future, this could be useful for electromagnetic transient simulations [42, 43-45].

It has mentioned that the modal transformation matrices obtained from the proposed correction procedure could be the base for simplified methods of electromagnetic transient analyses and simulations [42]. These simplifications should be checked using frequency scan and error analyses. Time domain analyses and transient simulations could be also applied for investigating the accuracy and the efficiency of the simplified modal transformation matrix applications [39-41].

Modal transformation matrices with constant and real elements have been investigated for application in systems with parallel three-phase circuits [46-47]. The correction procedure should be also adjusted for these systems, extending for different systems the mathematical development applied to three-phase transmission lines in this chapter [42, 44, 46-47].

VII. CONCLUSIONS

Based on the exact modal transformations, we suggested some applications of this concept for analyses related to transmission lines. We presented an alternative method for calculating the transmission line parameters from impedance values. It can be considered a theoretical method that is the base for the determination of an equivalent conductor from a bundled conductor. This is the second application shown in this chapter and it used the exact modal transformation matrices for calculating the parameters of the equivalent conductor. For typical symmetrical three-phase transmission line, the exact transformation matrices were separated into two matrices. One of them is a constant real matrix. This technique reduces the complex elements in the obtained transformation matrices, simplifying the numeric routine where these matrices are used.

Considering some approximations for the exact modal transformations, the eigenvector matrices are changed into Clarke's matrix for three typical three-phase transmission line cases. One of these is a symmetrical line and the others are asymmetrical lines. We analyzed the errors of these changes and concluded that the asymmetrical geometrical line characteristics can increase these errors. Because of this, we suggested a correction procedure application to Clarke's matrix. Based on Clarke's matrix, the new transformation matrices have small imaginary parts and their elements are smoothly influenced by the frequency. These characteristics are interesting because from these new transformation matrices, some simplifications can be obtained. For example, only the real part of the new transformation matrices can be used for applications to the EMTP programs.

REFERENCES

[1]   Dommel, HW. *Electromagnetic Transients Program-Rule, Book*, Oregon, 1984.
[2]   Microtran Power System Analysis Corporation, *Transients Analysis Program Reference*

Manual, Vancouver, Canada, 1992.

[3]   Morched, A; Gustavsen, B; Tartibi, M. "A Universal Model for Accurate Calculation of Electromagnetic Transients on Overhead Lines and Underground Cables", *IEEE Trans. on Power Delivery*, vol. 14, no. 3, 1032-1038, July 1999.

[4]   Brandão Faria, JA. "Overhead Three-phase Transmission Lines – Non-diagonalizable situations", *IEEE Transactions on Power Delivery*, vol. 3, no. 4, October 1988.

[5]   Brandão Faria, JA; Briceño Mendez, J. "Modal Analysis of Untransposed Bilateral Three-phase Lines - a Perturbation Approach", *IEEE Transactions on Power Delivery*, vol. 12, no. 1, January 1997.

[6]   Brandão Faria, JA. Briceño Mendez, J. "On the Modal Analysis of Asymmetrical Three-phase Transmission Lines using Standard Transformation Matrices", *IEEE Transactions on Power Delivery*, vol. 12, no. 4, October 1997.

[7]   Clarke, E. *Circuit Analysis of AC Power Systems*, vol. I, Wiley, New York, 1950.

[8]   Prado, AJ; Pissolato Filho, J; Kurokawa, S; Bovolato, LF. "Eigenvalue Analyses of Two Parallel Lines using a Single Real Transformation Matrix", *The 2005 IEEE/Power Engineering Society General Meeting*, CD-ROM, 12-16 June 2005, San Francisco, USA.

[9]   Prado, AJ; Pissolato Filho, J; Kurokawa, S; Bovolato, LF. "Non-transposed three-phase line analyses with a single real transformation matrix", *The 2005 IEEE/Power Engineering Society General Meeting*, CD-ROM, 12-16 June 2005, San Francisco, USA.

[10] Prado, AJ; Pissolato Filho, J; Kurokawa, S; Bovolato, LF. "Transmission line analyses with a single real transformation matrix - non-symmetrical and non-transposed cases", The 6th Conference on Power Systems Transients (IPST'05), CD-ROM, 19-23 June 2005, Montreal, Canada.

[11] Wedepohl, LM; Nguyen, HV; Irwin, GD. "Frequency –dependent transformation matrices for untransposed transmission lines using Newton-Raphson method", *IEEE Trans. on Power Systems*, vol. 11, no. 3, 1538-1546, August 1996.

[12] Nguyen, TT; Chan, HY. "Evaluation of modal transformation matrices for overhead transmission lines and underground cables by optimization method", *IEEE Trans. on Power Delivery*, vol. 17, no. 1, January 2002.

[13] Nobre, D.M; Boaventura, WC; Neves, WLA. "Phase-Domain Network Equivalents for Electromagnetic Transient Studies", *The 2005 IEEE Power Engineering Society General Meeting*, 12-16 June 2005, CD-ROM, San Francisco, USA.

[14] Budner, A. "Introduction of Frequency Dependent Transmission Line Parameters into an Electromagnetic Transients Program", *IEEE Trans. on Power Apparatus and Systems*, Vol. PAS-89, 88-97, January 1970.

[15] Carneiro Jr., S; Martí, JR; Dommel, HW; Barros, HM. "An Efficient Procedure for the Implementation of Corona Models in Electromagnetic Transients Programs", *IEEE Transactions on Power Delivery*, vol. 9, no. 2, April 1994.

[16] Martins, TFRD; Lima, ACS; Carneiro Jr., S. "Effect of Impedance Approximate Formulae on Frequency Dependence Realization", The 2005 *IEEE Power Engineering Society General Meeting*, 12-16 June 2005, CD-ROM, San Francisco, USA.

[17] Marti, JR. "Accurate modelling of frequency-dependent transmission lines in electromagnetic transients simulations", *IEEE Trans. on PAS*, vol. 101,. 147-155, January 1982.

[18]   Wedepohl, LM. "Application of Matrix Methods to the Solution of Travelling-wave Phenomena in Polyphase Systems", *Proceedings IEE,* vol. 110,. 2200-2212. December, 1963.

[19]   Wedepohl, LM; Wilcox, DJ. "Transient analysis of underground power-transmission system–system model and wave propagation characteristics", *Proceedings of IEE,* vol. 120, no. 2, 253-260, 1973.

[20]   Kurokawa, S; Pissolato, J; Tavares, MC; Portela, CM; Prado, AJ. "A new procedure to derive transmission-line parameters: applications and restrictions", *IEEE Trans. on Power Delivery,* vol. 21, no. 1, 492-498, January, 2006.

[21]   Hofmann, L. "Series expansions for line series impedances considering different specific resistances, magnetic permeabilities, and dielectric permittivities of conductors, air, and ground", *IEEE Trans. on Power Delivery,* vol. 18, no 2, 564-570, Apr. 2003.

[22]   Portela, C; Tavares, MC. "Modeling, simulation and optimization of transmission lines. Applicability and limitations of some used procedures", *IEEE PES Transmission and Distribution,* 2002, São Paulo, Brazil, 2002.

[23]   Semlyen, A. "Some frequency domain aspects of wave propagation on nonuniform lines", *IEEE Trans. on Power Delivery,* vol. 18, no 1, 315-322, Jan. 2003.

[24]   Akke, M; Biro, T. "Measurements of the frequency-dependent impedance of a thin wire with ground return", *IEEE Trans. on Power Delivery,* (Digital object identifier 101109/TPWRD.2004.834320).

[25]   Koolár, LE; Farzaneh, M. "Vibration of bundled conductors following ice shedding", *IEEE Trans. Power Delivery,* vol. 11, no 2, 2198-2206, April 2008.

[26]   Adams, GE. "An analysis of the radio-interference characteristics of Bundled Conductors", *AIEE Trans. Power Apparatus and Systems,* vol. 75, no.3, pp. 1569-1584, 1957.

[27]   Trinh, NG; Vincent, C. "Bundled-conductors for EHV transmission systems with compressed SF6 insulation", AIEE Trans. Power Apparatus and Systems, vol. 75, no 6, 2198-2206, 1978.

[28]   Dan, VV. "A rational choice of bundle conductors configuration", in *Proc.1998 Int. Symp. on Electrical Insulating Materials Conf.,* Toyohashi, Japan, 349-354.

[29]   Watson, N; Arrilaga, J. *Power Systems Electromagnetic Transients Simulation,* London: Institution of Electrical Engineers, 2003, 140-142.

[30]   Tu, VP; Tlusty, J. "The calculated methods of a frequency-dependent series impedance matrix of overhead transmission lines with a lossy ground for transient analysis problem", in Proc. 2003 Large Engineering Systems Conference on Power Engineering, Montreal, Canada, 159-163.

[31]   Nayak, RN; Sehgal, YK. Sen, S. "EHV transmission line capacity enhancement through increase in surge impedance loading level", in *Proc. 2006 IEEE Power India* Conference, New Delhi.

[32]   Martinez, JA; Gustavsen, B; Durbak, D. "Parameters determination for modeling system transients - Part I: overhead lines", *IEEE Trans. Power Delivery,* vol. 20, no. 3, 2038-2044, July 2005.

[33]   Mingli, W; Yu, F. "Numerical calculations of internal impedance of solid and tubular cylindrical conductors under large parameters", *IEE Proc. Generation, . Transmission and Distribution,* vol. 151, no 1, 67-72, 2004.

[34]   Tu, VP; Tlusty, J. "The calculated methods of a frequency-dependent series impedance

matrix of a overhead transmission line with a lossy ground for transient analysis problem", in *Proc. 2003 Large Engineering Systems Conference on Power Engineering*, Montreal, Canada, 154-158.

[35] Sinclair, AJ. Ferreira, JA. "Analysis and design of transmission line structures by means of the geometric mean distance", in Proc. 2006 IEEE Africon Conference in Africa, Stellenbosch, South Africa, 1062-1065.

[36] Kurokawa, S; Costa, ECM; Pissolato, J; Prado, AJ; Bovolato, LF. "An alternative model for bundled conductors considering the distribution of the current among the subconductors", *IEEE Latin America Transactions*, in press.

[37] Kurokawa, S; Daltin, RS; Prado, AJ; Pissolato, J. "An alternative modal representation of a symmetrical nontransposed three-phase transmission line", *IEEE Trans. on Power Systems*, vol. 22, no. 1, 500-501, February, 2007.

[38] Fernandes, AB; Neves, LA. "Phase-domain transmission line models considering frequency-dependent transformation matrices", *IEEE Transactions on Power Delivery*, vol. 19, No 2, 708-714, April 2004.

[39] Prado, AJ; Kurokawa, S; Pissolato Filho, J; LF. Bovolato, "Asymmetric transmission line analyses based on a constant transformation matrix", The 2008 *IEEE/PES Transmission and Distribution Conference and Exposition: Latin America*, 13-15 August 2008, CD-ROM, Bogotá, Colombia.

[40] Prado, AJ; Kurokawa, S; Pissolato Fiho, J; Bovolato, LF. "Single real transformation matrix application for asymmetrical three-phase line transient analyses", *The 2008 IEEE/PES Transmission and Distribution Conference and Exposition: Latin America*, 13-15 August 2008, CD ROM, Bogotá, Colombia.

[41] Prado, AJ; Pissolato Fiho, J; Kurokawa, S; Bovolato, LF. *"Clarke s matrix correction procedure for non transposed three-phase transmission lines"*. The 2008 IEEE/PES General Meeting, 20-24 July 2008, CD ROM, Pittsburgh, Pennsylvania, USA.

[42] Prado, AJ; Kurokawa, S; Pissolato Filho, J; Bovolato, LF. "Voltage and current mode vector analyses of correction procedure application to Clarke's matrix - symmetrical three-phase cases", *Journal of Electromagnetic Analysis and Applications*, vol. 2, no. 1, 12, March, 2010.

[43] Kurokawa, S; Prado, AJ; Pissolato Filho, J; Bovolato, LF; Daltin, RS. "Alternative proposal for modal representation of a non-transposed three-phase transmission line with a vertical symmetry plane", *IEEE Latin America Transactions*, vol. 7, no. 2, 182-189, June, 2009.

[44] Kurokawa, S; Yamanaka, FNR; Prado, AJ; Pissolato Filho, J. "Inclusion of the frequency effect in the lumped parameters transmission line model: state space formulation", *Electric Power Systems Research*, vol. 79, no. 7, 1155-1163, July, 2009.

[45] Kurokawa, S; Daltin, RS; Prado, AJ; Pissolato Filho, J. "An alternative modal representation of a symmetrical non-transposed three-phase transmission line", *IEEE Transactions on Power Systems*, vol. 22, no. 1, 500-501, February, 2007.

[46] Campos, JCC; Pissolato Filho, J; Prado, AJ; Kurokawa, S. "Single Real Transformation Matrices Applied to Double Three-phase Transmission Lines", *Electric Power Systems Research*, vol. 78, no. 10. 1719-1725, October, 2008.

[47] Prado, AJ; Pissolato Filho, J; Kurokawa, S; Bovolato, LF. "Modal transformation analyses for double three-phase transmission lines", *IEEE Transactions on Power Delivery*, vol. 22, no. 3, 1926-1936, July, 2007.

In: Electric Power Systems in Transition
Editors: Olivia E. Robinson, pp. 75-136

ISBN: 978-1-61668-985-8
© 2010 Nova Science Publishers, Inc.

Chapter 2

LOAD MODELING IN POWER SYSTEMS: INDUCTION MOTORS

Carlos F. Moyano and Gerard Ledwich

Queensland University of Technology – QUT, 2
George Street – Brisbane (4000) – QLD, Australia

ABSTRACT

Modelling the power systems load is a challenge since the load level and composition varies with time. An accurate load model is important because there is a substantial component of load dynamics in the frequency range relevant to system stability. The composition of loads need to be characterized because the time constants of composite loads affect the damping contributions of the loads to power system oscillations, and their effects vary with the time of the day, depending on the mix of motors loads.

This chapter has two main objectives: 1) describe the load modeling in small signal using on-line measurements; and 2) present a new approach to develop models that reflect the load response to large disturbances.

Small signal load characterization based on on-line measurements allows predicting the composition of load with improved accuracy compared with post-mortem or classical load models. Rather than a generic dynamic model for small signal modelling of the load, an explicit induction motor model is used so the performance for larger disturbances can be more reliably inferred. The relation between power and frequency/voltage can be explicitly formulated and the contribution of induction motors extracted. One of the main features of this work is the induction motor component can be associated to nominal powers or equivalent motors.

When large disturbances are considered, loads play a major role in the recovery of the voltage. Among the different types of loads, induction motors have a extremely non-linear characteristic. After a disturbance, an induction motor can be reaccelerated, stalled or tripped. The stalled case is the most important case because of the large amount of reactive power demanded, which can influence the recovery times of voltage after the fault is cleared and has the potential to drive the system to voltage collapse. Regarding the importance of induction motors and the characterization of the pre-fault state using on-line measurements, a new model for large disturbance is proposed in this work. It will be shown that information contained in the on-line measurements allows description of the static and induction motor components in the pre-fault state. The induction motor component can be described as groups of motors with similar inertia and nominal power, which can be obtained from the small signal model. To obtain a complete characterization of the response of the load to large disturbances, two indices are proposed in this work: the severity index and the tripping index. The first one will describe the probability of motors in a group stall after the fault and the second describes the amount of load that will be lost in a major disturbance. Both indexes are directly associated with the recovery of the voltage. The importance of the indices is that they can be identified through a limited set of fault measurements while a full response model across the full range of depth and duration of disturbances would take hundreds of fault records.

1. INTRODUCTION

The current industry approach is strongly based on simulation of power systems to enable the assessment of possible technical solutions and operational practices. A proper representation of the power system components is necessary to replicate correctly the system behaviour and avoid over/under building of the system, and in extreme cases can lead to degradation of reliability [1].

Simulations of Power Systems can be classified into static and dynamic studies. Static studies include the power flow, optimal power flow and other studies where the objective is to obtain a snapshot of the state of the system regardless the transient behaviour. Dynamic studies include transient stability, fast voltage collapse and small signal stability. All these studies involve transient simulations.

Detailed and accurate models for Generators and Transmission lines are available for both types of studies, but load modelling is still an open field, especially in dynamic load model representation [1].

Static studies of power systems such as power flow and optimal power flow usually consider all the loads at low voltage buses lumped at the high voltage buses and the dependence of loads to voltage changes is usually not considered because of the action of the tap changers. In long term voltage stability studies based on power flow, this is not accurate enough and the load has to be modelled with a higher degree of detail or taken from dynamic simulation models [1].

The necessity of an accurate representation of the loads in dynamic studies has been known for a long time. Reports from IEEE Task Force on Load Representation for Dynamic Performance [1, 2]) and other groups as EPRI [3] and many conference and journal papers highlighted the importance of dynamic load modelling.

Accurate load modelling is a difficult task due to several factors such as [2],

a) large numbers of diverse load components;
b) ownership and location of load devices in customer facilities is not directly accessible to the electric utilities;
c) changing load composition with time of day and week, seasons, weather;
d) lack of precise information on the composition of the load

e) uncertainties regarding the characteristics of many load components, particularly for large frequency or voltage variations

The large number and diversity of loads in power systems make it extremely difficult to model each and every load. To overcome this problem, aggregated load models with different load components are normally used for stability studies.

The IEEE Task force characterized multiple types of loads connected to a bus. Load types considered by the IEEE task force are static, induction motors, synchronous motor and transformer saturation. Each individual load type may have multiple representations. A static load model expresses the characteristics of the load at any instant of time as a polynomial or some other algebraic functions such as exponential function. This representation is based on the voltage and frequency dependency of the load observed within a limited range of voltage and frequency variations [4]. The voltage and frequency dependency of load characteristics has been represented by the exponential model [1]:

$$\frac{P}{P_{frac}P_0} = K_{pz}\left(\frac{V}{V_0}\right)^2 + K_{pi}\frac{V}{V_0} + K_{pc} + K_{p1}\left(\frac{V}{V_0}\right)^{n_{pv1}}(1+n_{pf1}\Delta f) + K_{p2}\left(\frac{V}{V_0}\right)^{n_{pv2}}(1+n_{pf2}\Delta f) \tag{1}$$

$$K_{pz} = 1 - (K_{pi} + K_{pc} + K_{p1} + K_{p2}) \tag{2}$$

where P_{frac} is the fraction of the bus represented by the static model.

$$\frac{Q}{Q_{frac}Q_0} = K_{qz}\left(\frac{V}{V_0}\right)^2 + K_{qi}\frac{V}{V_0} + K_{qc} + K_{q1}\left(\frac{V}{V_0}\right)^{n_{qv1}}(1+n_{qf1}\Delta f) + K_{q2}\left(\frac{V}{V_0}\right)^{n_{qv2}}(1+n_{qf2}\Delta f) \tag{3}$$

$$Q_0 \neq 0 \tag{4}$$

$$K_{qz} = 1 - (K_{qi} + K_{qc} + K_{q1} + K_{q2}) \tag{5}$$

where Q_{frac} is the fraction of the bus represented by the static model. In the previous equations Δf is the change in frequency, and the coefficients K are the parameters of the model. P and Q are the active and reactive components of the load when the bus voltage magnitude is V. The subscript '0' identifies the values of the respective variables at the initial operating condition. The exponents 0, 1, or 2 of the model for the voltage terms which are not affected by frequency changes represent constant power, constant current or constant impedance characteristics, respectively. For composite loads, their values depend on the aggregate characteristics of load components. The response of most of the composite loads to voltage and frequency changes is fast at least for modest amplitudes of voltage and frequency change. The use of static models is justified in such cases [5].

Synchronous motors and transformer saturation have appropriate models that are available in [1] but are out of the scope of this work.

The work is to characterize the dynamic response of key load components, one of the critical loads are the induction motors. Induction motors contribute large portion of total load in any power system [6]. They are one of the most important components of the load, with a highly nonlinear behaviour, presenting very different responses if the motor is in normal operation, tripping or stalling [7].

Problems that highlighted the importance of modelling induction motors are the delay of the transmission voltage recovery [8] and the influence of induction motors in small signal stability.

Lack of adequate dynamic motor models is suspected to be the major source of discrepancies between field measurements and large scale simulations [1]. Induction motors and the way they contribute to power systems stability have been researched for a long time [9]. Early in 1982 the authors of [10] pointed out the necessity of representing major blocks of induction motor load by dynamic models including both inertial and rotor flux dynamics. The research carried out by Western Systems Coordinating Council (WSCC) concluded that the single most important sensitivity was the percentage of motors modelled at the load bus. A level of 20 to 30% of motor load provided the best simulated of the phenomena described in that case. Varying the motor inertia and impedances had varying impacts but, relatively, not as great as the motor percentages [11].

1.1. Approaches to Obtain a Model of the Load

The characteristics of any power system load can be determined either by using component based approach or/and measurement based approach. The component based approach focuses on the categorization of lumped load (at a bulk power delivery point) into various load classes such as residential, commercial and industrial. Each class of load would be further divided into different load components such as lighting, heating, motors, etc. The electrical characteristics of each load component would be analysed before aggregate modelling. The EPRI Load Model Synthesis Program Package (LOADSYN) converts load mix, load composition, and load characteristic data into the parameters required for power flow and transient stability programs [12]. An investigation of several methods for aggregating induction motors in the LOADSYN program is described in [13].

The measurement-based approach involves direct measurements at representative substations and feeders to determine the voltage and frequency sensitivity of the active P and reactive Q load. The instrumentation requirements for monitoring the dynamic performance of electric power systems are reviewed in [3, 13, 31]. The data is obtained from measurements in-situ, and includes voltage and frequency variations, and the corresponding variations in active and reactive load, either to intentional disturbances, test measurements, natural events and normal operation data. By fitting the measured data to a model, the parameters of the load model are identified [15] provided that the spectrum of the disturbances is spectrally rich.

1.2. Objectives of this Work

The first objective of this work is to show how measurements of a composite load response to normal power system variations can be used to continuously identify the load dynamic parameters. In this work the identification is reduced to the induction motor component of the load. It will be shown that the techniques developed allow characterizing the induction motors presented in real data.

The second objective is to shown that pre-fault information from the small signal model and simulated response from induction motors can be suitable to create a methodology that considerably reduces the number of disturbances required to test the model. This work presents the formulation of two indexes that characterize the effects of large disturbances on induction motors, the severity index and the tripping index. Using the information available from the small signal model and the two indexes, a model that can reproduce the aggregate response of a group of induction motors to large disturbance can be obtained.

A brief description of induction motors is presented in the next section, and its findings are used to justify the proposed techniques.

2. INDUCTION MOTORS

2.1. Models for Induction Motor Simulation

Induction motors can be classified by their construction as wound and squirrel cage types. The most widely used motor is the squirrel cage type, which can be classified as single cage, double cage and deep bar rotor cage. In [16] Parks' equation in the synchronous reference of the double cage induction motor are represented in equation (6)-(7), where p is the derivative operator, $\wp$ is the number of pairs of poles and J is the combined inertia of the motor and load. The coefficients R_s, R_{rl}, R_{r2}, L_s, L_{rl}, L_{r2} and M are calculated from the steady state parameters.

$$\begin{bmatrix} v_{sd} \\ v_{sq} \\ 0 \\ 0 \\ 0 \\ 0 \end{bmatrix} = \begin{bmatrix} R_s+pL_s & -\omega L_s & pM & -\omega M & pM & -\omega M \\ \omega L_s & R_s+pL_s & \omega M & pM & \omega M & pM \\ pM & -\omega sM & R_{r1}+pL_{r1} & -\omega sL_{r1} & pM & -\omega sM \\ \omega sM & pM & \omega sL_{r1} & R_{r1}+pL_{r1} & \omega sM & pM \\ pM & -\omega sM & pM & -\omega sM & R_{r2}+pL_{r2} & \omega sL_{r2} \\ \omega sM & pM & \omega sM & pM & \omega sL_{r2} & R_{r2}+pL_{r2} \end{bmatrix} * \begin{bmatrix} i_{sd} \\ i_{sq} \\ i_{r1d} \\ i_{r1q} \\ i_{r2d} \\ i_{r2q} \end{bmatrix} \quad (6)$$

$$p\omega_m = \frac{1}{J}(T_e - T_r) \quad (7)$$

The slip is defined by

$$s = \frac{(\omega - \wp \omega_m)}{\omega}; \quad \omega = 2\pi f \quad (8)$$

where ω_m is the motor speed and f is the frequency of the system.

The motor torque is defined using a normalized Parks' transformation as

$$T_e = \wp M (i_{sq}(i_{r1d} + i_{r2d}) - i_{sd}(i_{r1q} + i_{r2q})) \tag{9}$$

And the resistive torque is defined by

$$T_r = K_0 + k_1 \omega_m + k_2 \omega_m^2 \tag{10}$$

where the coefficients depend on the type of load: constant load K_0, linear load k_1 and quadratic load k_2. The inertia J usually is defined through the inertia constant H

$$H = \frac{J(2\pi f)^2}{2\wp^2 P} \tag{11}$$

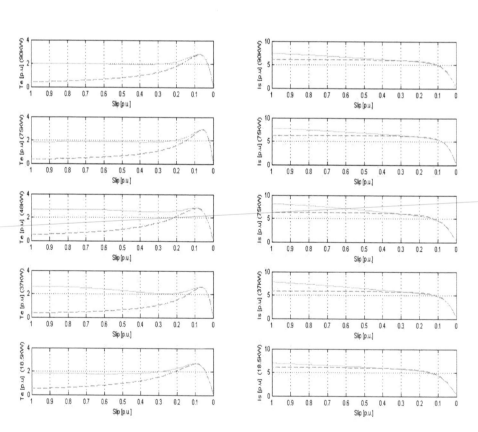

Figure 1. Torque and current characteristics for different motors

Equations (6)-(11) describe the induction motor model of a double cage type, which can also be used to describe the behaviour of deep bar induction motors. The induction motor simulation requires the determination of the steady state or initial values of the voltages, currents, and slip. Once the initial values are known, the differential equations describing the change on the values of the variables with respect to time are solved. The new values for the currents are used to recompute the values of the rotor voltages, currents, and powers at each step.

The common models used to simulate the induction motor behaviour are obtained simplifying the model presented in Equations (6) to (11).

The fifth order model presented in [17] for a single cage induction motor is obtained from the previous set of equations considering the currents and derivatives associated with the inner cage i_{r1d} and i_{r1q} to be zero.

The order of the induction machine model can be reduced from five to three by omitting the stator flux transients. This third-order model is often referred to as the transient stability model or the neglecting stator transient's model. The derivatives of the stator variables are set to zero in (6), and the stator variables are then solved as functions of the rotor currents and the rotor speed [18].

First-Order Model equations of the machine are obtained by setting all the derivatives in (6) to zero. Combining these steady-state equations with the equation of motion (7) creates a first-order model of the induction machine, where the only state variable is the rotor speed [18].

2.2. Characteristics Speed-Torque and Current Torque

Single and double cage present different torque-speed and current-speed characteristics. The presence of a double cage modifies the torque-speed characteristic and provides a higher initial torque. Also the initial current is greater in the double cage case. Figure 1 from [19] shows the different torque and current characteristics as a function of the slip for different motors with nominal power ranging from 18.5kW to 90 kW.

In the previous figure, different nominal power motors with both single and double cage characteristic are shown. The dotted lines represent the single cage and the smooth line the double cage. One observation is the maximum torque for both models are similar and can be considered as the same for power system simulation purposes. The starting torque is clearly not the same. Double cage motors have a more powerful starting torque and their starting currents are higher than the single cage model.

The equivalent circuits for single and double cage are shown in the next figure [20],

The difference in torque-speed and current-speed characteristics can be explained using the models shown in figure 2. The inner cage in the double cage motor provides the source of torque for low frequency currents in the rotor. The outer cage, which is similar to the single cage model, provides the torque when the motor speed is close to the nominal speed. This is shown in figure 3 derived from [16].

Parameters of the double cage model can be obtained as explained in [19, 20], where the following Tables 1 and 2 were extracted. In these Tables the nominal power P_k is expressed in kW. Table 1 shows the coefficients that can be used to calculate an approximation of the double cage induction motor parameters.

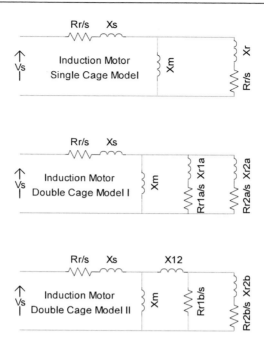

Figure 2. Single and double cage models from the steady state parameters.

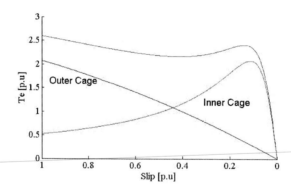

Figure 3. Effect of the inner and outer cages in the double cage model.

Parameters of the single cage model can be obtained as shown in Table 2 [20]. As in Table 1 the nominal power P_k is in kW

The torque characteristic for points near the normal operation point can be considered similar in single cage or double cage. Note that the normal operation point is under the maximum torque or breaking torque point for most of the applications.

The maximum torque of an induction motor modelled with the single cage can be represented as function of its parameters [21]. In the double cage model there is not analytical expression for the maximum torque [20]. Regarding the single cage induction motor model, the breakdown or maximum torque for this model (at $s=s_{max}$) is

$$T(s_{max}) = \frac{3}{2\omega_s} * V_1^2 * \left(\frac{X_m^2}{R_1^2 + (X_1 + X_m)^2} \right) * \frac{R_1^2 + (X_1 + X_m)^2}{R_1 X_m^2 + \sqrt{R_1^2 X_m^4 + (X_m(R_1^2 + X_1^2 + X_1 X_m) + X_{20})}} \quad (12)$$

Equation (12) shows a dependency of the maximum torque as a function of the square of the input voltage and the parameters of the motor. The slip where the maximum torque happen is

$$s_{max} = \frac{R_{20}}{\sqrt{R_1^2 X_m^4 + (X_m(R_1^2 + X_1^2 + X_1 X_m) + X_{20})}}$$ (13)

Table 3 shows the relation between the minimum allowable torque ratios of NEMA design A and B motors [21]. Similar characteristics can be found in other norms.

Table 1. Parameters of the double cage model.

R_s	$0.0362P_k^{-0.3926}$
R_{r1}	$0.0713P_k^{-0.1315}$
R_{r2}	$0.1090P_k^{-0.4392}$
X_M	$1.2609P_k^{0.1277}$
X_s	$0.0519P_k^{0.0533}$
X_{r1}	$0.0379P_k^{0.0323}$
X_{r2}	$0.1606P_K^{-0.0837}$

Table 2. Parameters of Single Cage model.

R_s	$0.0264P_k^{-0.3453}$
R_r	$0.0461P_k^{-0.3255}$
X_M	$1.4209P_k^{0.0829}$
X_r	$0.0625P_k^{0.0572}$
X_s	$0.0625P_k^{0.0572}$

2.3. Influence of the Motor Inertia

The inertia constant of induction motors affect small-signal stability, especially affecting damping. The influence depends on the system structure, the parameters of components, and the initial operating condition of the system. Induction motor loads mainly influence the inter-area oscillation mode in multi-machine system. The extent and the trend of the influence are related to inertia, the fraction of power related to induction motors and the location of them [22].

Figure 4 shows the transfer function of a 630KW motor when the inertia is varied. Higher inertia means a displacement to the left in that figure.

**Table 3. Minimum Allowable Torque Ratios of
NEMA Design A and B Motors.**

HP	T_{st}/T_{fl}	T_{max}/T_{fl}	T_{st}/T_{max}
10 - 75	1.05-1.65	2.00	0.525-0.825
100-200	1.00-1.25	2.00	0.500-0.625
250-450	0.70-1.00	1.75	0.400-0.571

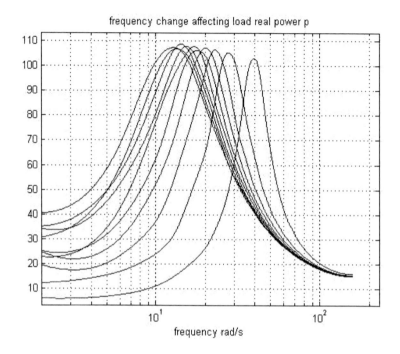

Figure 4. Different inertias for a 630KW motor

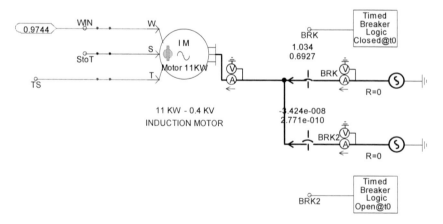

Figure 5. 11 KW Induction motor with ideal source.

The induction motors inertia also plays an important role when major disturbances are simulated. If the inertia of the motor and the mechanical load connected to the shaft is small, the motor deceleration is faster than with a greater inertia. When the voltage tries to recover, the motor will draw a greater amount of current. The next figures illustrate this matter, considering the effect of voltage sags on the induction motors response

Figure 5 shows an 11kW induction motor fed by an ideal source as modelled in PSCAD [32]. The only resistance between the ideal source and the induction motor is the breaker with a value of *0.01* ohms. At *t=0.0* sec the breaker BRK is closed and at *t=5.0* sec BRK is open and then the breaker BRK2 is closed for *0.5* sec, connecting to a second ideal source with *40%* of the original voltage. At *t=5.5* sec the situation is reversed and the original voltage is re-established.

Figure 6 and Figure 7 show the simulation for two different values of inertia. In the first case the inertia of the motor and mechanical load is *0.36* sec, which is twice the inertia presented in a manufacturer catalogue [23]. These values can be considered suitable to simulate a set motor-compressor for an 11kW motor working close to *1500*rpm. The second case shows the same simulation but the inertia is now *1.5* sec, which could be related to other type of load. In both cases the motors are working with constant torque control at the time of the disturbance.

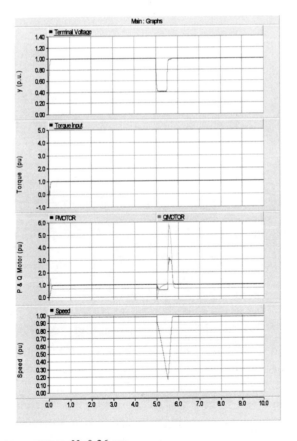

Figure 6. Induction motor response. H=0.36 sec.

Figure 6 and Figure 7 show the deceleration of the motors is strongly influenced by the value of the inertia constant. Note the differences in speeds and reactive power demanded.

The deceleration of the motor of Figure 6 is stronger than the other case. When the voltage is re-established, the motor in Figure 6 needs to reaccelerate from a very low speed, demanding a current value close to the starting current. This difference is the cause of a higher recovery time after the fault and a greater demand of active and reactive power from the grid for the motor in Figure 6.

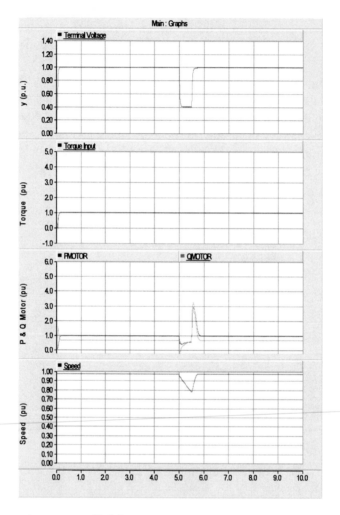

Figure 7. Induction motor response. H=1.5 sec.

The next simulation shows the effects of an increase in the resistance of the breaker to 20 times that of the motor with inertia constant equal to 0.36sec. The logic of breakers is the same as in the previous case.

As can it be seen in Figure 8 the motor takes a longer time to recover the pre-disturbance state and the voltage at the motors' terminals is depressed by the effect of a strong resistance between the ideal source and the induction motor. If the inertia constant of the motor is set at 1.5 sec the results are similar, but the recovery of the motor requires less time and the voltage at the terminal is less depressed. The peak of the active and reactive power drained by the motor is lower in this case, but the recovery time is greater. The consequence of this for load modelling aggregate loads is that the inertia and impedance of the aggregate motor is of key importance.

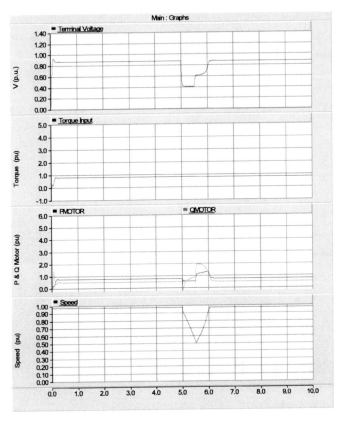

Figure 8. Induction motor response with the impedance of the source increased.

2.4. Small Motors

Low values of inertia mean a longer time of recovery of the speed after a voltage sag. Since small motors have small inertia on a p.u. basis, the smaller motors would be more affected by disturbances. Motors used in small residential air-conditioners (central and window types) and refrigerators tend to stall when the voltage is reduced below 60% for 5 cycles or longer [8]. Stalling occurs because the torque produced by the motor under reduced-voltage conditions cannot overcome the back pressure of the compressor due to reduced torque under long times of system voltage recovery after a fault. Under these conditions, the motor will continue to draw very high current, depending on the voltage level, approaching the locked-rotor current at rated voltage. The motor will normally trip on thermal overload after 3 to 30 seconds. But until it trips, it will draw a considerable amount of active power and an even-higher amount of reactive power from the system since the power factor of the motor is also significantly reduced under these conditions. This "sustained" increase in active and reactive power demand would slow voltage recovery even more, and can even cause voltage collapse in certain situations. Larger commercial and industrial motors are typically equipped with an under voltage protection relay that trips-off the unit for voltages less than 70% in 0.1 seconds [25]. Thus, commercial and industrial motors do not affect the voltage recovery as adversely as residential air conditioners under the same conditions [26].

"Prone-to-stall" motor loads, particularly residential air-conditioners and refrigeration motor loads, can significantly impact system stability following a major disturbance. Dynamic simulation models utilizing dynamic motor models that do not represent stalled motor conditions will yield voltage recovery times that may be faster than actual [27].

This section presented a brief explanation of the induction motor characteristics and two major findings: 1) if the nominal power of the induction machine is available, then the parameters can be calculated as a function of the power, and 2) the combined inertia of the induction motors and mechanical load influences the behaviour of the induction motor under major disturbances. Small inertia motors, particularly with compressor loads, are prone to stall under a voltage dip.

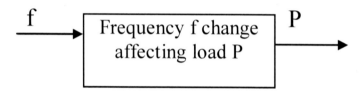

Figure 9. Basic Model.

3. MEASUREMENT BASED LOAD MODEL - SMALL SIGNAL

3.1. Analysis Approach

Loads related to induction motors can be separated considering that if the system frequency f were varied but all voltage magnitudes were held constant then just the induction motors would show transient and steady state change in active power P. Measurements of the observed frequency perturbations at the supplying bus and the variations of P and Q to the load are available using PMU (Phase Measurement Units) [31]. In this work, all reference to measurements means they are time synchronized using GPS.

The correlation of these measurements allows inferring a transfer function from frequency to load power. This approach is similar to those where a step in voltage allows observing the changes in load power.

Loads can be modelled as fractions of motor load plus static loads described by equations 1 to 5. But there is also a continual variation of the demand by customers switching loads ON and OFF that must be considered. The switching changes in customer load will cause changes in the angle and voltage magnitude at the supplied bus. These angle changes will influence the measurements of supply bus frequency. It is well known that large changes of customer load cause system frequency to drop as well as excite inter-area modes. Similarly an increase of customer load will cause a drop in the bus voltage. From these facts, it is inferred that the load can also have an effect on the power system and this effect is represented by a feedback term.

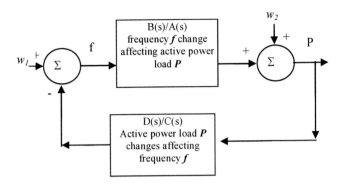

Figure 10. Feedback model.

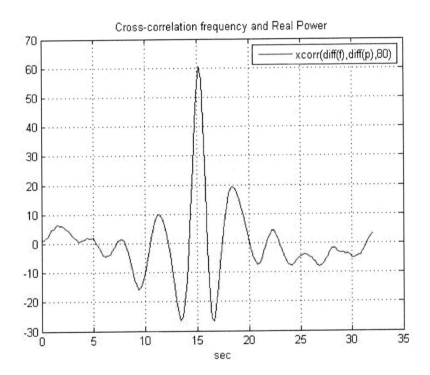

Figure 11. Cross-correlation between frequency and real power.

Regarding the feedback term in Figure 10, the identified transfer function from f to P could be in error if the feedback nature of the system is ignored. The evidence for the feedback structure is the non-causal correlation between f and P. Field measurements taken by QUT have shown a correlation in both a positive and negative time delay. This is seen in the cross correlation between the difference of frequency and the difference of power in Figure 11. This means an approach to disclose the feedback term is necessary.

One of the positive aspects for the identification of the f-P relation is that a very low value of the high frequency gain of the feedback term $D(s)/C(s)$ is expected. Because of generator inertia, the system frequency will not change rapidly to rapid changes of load power thus $(D(j\omega)/C(j\omega)) \to 0$ for $\omega \to \infty$. This means that observed high frequency variations in bus frequency can be identified with the term w_1 which is this case refers to the rapid changes in bus angle arising from the random changes in local load rather than the indirect effect via $D(s)/C(s)$.

Models to predict the next sample based on the history of f and P can be formed based on previous values. When f is dominated by the low pass filtered feedback signal, most of the measurement can be well predicted from previous measurements using tools such as least squares. If all the predictable portions of f are removed, the remaining unpredictable portion is a white noise signal with a flat spectrum [35]. This is a possible approach to extract the component causing the measurement variations. The unpredicted changes in frequency, associated with the white noise term w_1 in Figure 10 can be identified using this approach. Similarly the unpredicted changes in P (or Q signals) are associated with the term w_2 representing load changes.

Noise terms w_1 and w_2 can be used to form transfer functions from w_1 to X and w_1 to Y. The ratio of these terms identifies the feedforward portion of Figure 10. These transfer functions are identified using cross correlation approaches [29]. Here the transfer function is found as the spectrum of cross correlation of input to output divided by the autocorrelation of the input. The transfer function from w_2 to f and P can be found and the ratio used to identify the feedback term.

Consider variables $y(t)$ and $u(t)$ (output y and input u) with representation in the s and frequency domain $Y(s)$, $U(s)$, $Y(j\omega)$, $U(j\omega)$, where $(j\omega)$ means a fourier transformation of the sample data. Using the representation in s domain of the terms $A(s)$, $B(s)$, $C(s)$ and $D(s)$, the input and output can be expressed in the frequency domain as

$$Y(s) = \frac{B(s)}{A(s)} U(s) + W_2(s) \tag{14}$$

$$U(s) = \frac{D(s)}{C(s)} Y(s) + W_1(s) \tag{15}$$

Where $W_1(s)$ and $W_2(s)$ are the representations of the noise in the s domain Expressing U and Y from (14) and (15) as function of the noises yields

$$Y(s) = \left(\frac{B(s)}{A(s)} W_1(s) + W_2(s) \right) / \left(1 - \frac{B(s)}{A(s)} \frac{D(s)}{C(s)} \right) \tag{16a}$$

$$U(s) = \left(W_1(s) + \frac{D(s)}{C(s)} W_2(s) \right) / \left(1 - \frac{B(s)}{A(s)} \frac{D(s)}{C(s)} \right) \tag{16b}$$

Equations (16a)-(16b) show bias terms associated with the other input appearing in these equations.

Distributing the bias terms into the equations

$$Y(s) = \left(\frac{C(s)B(s)}{A(s)C(s) - B(s)D(s)} \right) W_1(s) + \left(\frac{A(s)C(s)}{A(s)C(s) - B(s)D(s)} \right) W_2(s) \tag{17}$$

$$U(s) = \left(\frac{A(s)C(s)}{A(s)C(s) - B(s)D(s)} \right) W_1(s) + \left(\frac{A(s)D(s)}{A(s)C(s) - B(s)D(s)} \right) W_2(s) \tag{18}$$

Calculating the cross-spectral density from input $U(s)$ and output $Y(s)$ to input noise $W_1(s)$ in the frequency domain, from [48],

$$\Phi_{YW_1}(j\omega) = E(W_1^*(j\omega)Y(j\omega)) =$$
$$E\left(\frac{C(j\omega)B(j\omega)}{A(j\omega)C(j\omega) - B(j\omega)D(j\omega)} \right) E(|W_1(j\omega)|^2) +$$
$$E\left(\frac{A(j\omega)C(j\omega)}{A(j\omega)C(j\omega) - B(j\omega)D(j\omega)} \right) E(W_1^*(j\omega)W_2(j\omega)) \tag{19}$$

Where E is the expectation operator. The second term of (19) is zero, since the noises are not correlated. $E(|W_1(j\omega)|^2)$ is the representation in frequency domain of the auto-correlation of $W_1(j\omega)$ term, and the cross-spectral density between $W_1(j\omega)$ and $Y(j\omega)$ can be written as

$$\Phi_{YW_1}(j\omega) = \frac{E(W_1^*(j\omega)Y(j\omega))}{E(|W_1(j\omega)|^2)} = E\left(\frac{C(j\omega)B(j\omega)}{A(j\omega)C(j\omega) - B(j\omega)D(j\omega)} \right) \tag{20a}$$

Following a similar development

$$\Phi_{UW_1}(j\omega) = \frac{E(W_1^*(j\omega)U(j\omega))}{E(|W_1(j\omega)|^2)} = E\left(\frac{A(j\omega)C(j\omega)}{A(j\omega)C(j\omega) - B(j\omega)D(j\omega)} \right) \tag{20b}$$

Defining $\xi(j\omega) = \dfrac{C(j\omega)}{A(j\omega)C(j\omega) - B(j\omega)D(j\omega)}$, and since A, B, C and D are supposed to be constant for each frequency value,

$$\Phi_{YW_1}(j\omega) = B(j\omega)\xi(j\omega)$$

$$\Phi_{UW_1}(j\omega) = A(j\omega)\xi(j\omega)$$

And thus the frequency response transfer function from input $u(t)$ to output $y(t)$ in the $\Phi_{UY}(j\omega)$ can be obtained through the noise $w_1(t)$

$$\Phi_{UY}(j\omega) = \frac{\Phi_{YW_1}(j\omega)}{\Phi_{UW_1}(j\omega)} = \frac{B(j\omega)}{A(j\omega)} \tag{21}$$

A similar development yields the frequency response transfer function of the feedback term using the output noise as

$$\Phi_{YU}(j\omega) = \frac{\Phi_{YW_2}(j\omega)}{\Phi_{UW_2}(j\omega)} = \frac{D(j\omega)}{C(j\omega)}$$

(22)

3.1.1. Examples

Consider a system where the feedforward term is first order with a pole at 0.5 and a feedback system with resonant poles at 1Hz and damping 0.05. The problems of direct identification are seen in Figure 12 which shows that just fitting a transfer function between u and y yields a model with poles at a combination of the feedforward and feedback locations. When the noise terms w_1 and w_2 are extracted we see a good match between identified and original systems in Figure 13. In Figure 13 ratio 1 and ratio 2 are the calculated magnitude and phase of the transfer function as explained before, and ref1 and 2 are the real magnitude and phase.

3.1.2. Identification with feedthrough

The next example is a case where the feedback term is a low pass filter. This can be considered similar to the feedback term of real power systems where changes in P would affect the frequency with a low pass characteristic. This assumption is based on the fact that power system frequency is driven mainly by the governor and electromechanical modes. If the feedback term had a portion of direct feedthrough then there will be a component of the white noise term w_2 in the measurement u as well as the w_1 term.

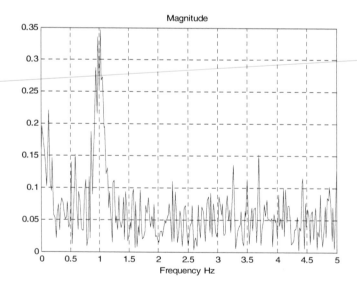

Figure 12. Direct transfer function estimate U-Y.

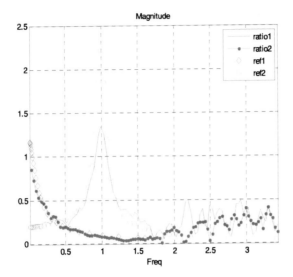

Figure 13. Quality of fit.

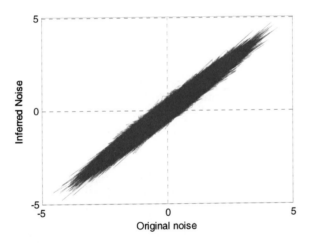

Figure 14. Quality of identification of the noise W1.

When a direct feedthrough term $B(s)=0.2+(0.6s+7.89)/(s^2+0.62s+39.5)$ is added to the previous example and the best prediction to u is applied to find the white noise term the estimate is no longer a good estimate of the w_1 sequence. Figure 14 shows the mapping between the identified signal and w_1 where the signal is no longer clean. The identified signal is a combination of w_1 and w_2 and can no longer be used to separate the feedforward and feedback components.

The process of identification of the separate feedforward and feedback components is more difficult when the feedthrough term is present. This characteristic could be expected in some voltage magnitude to active power relationships. High frequency variations in P are present in real measurements and high frequency variations being caused in the bus voltage magnitude are expected. Consider the case when both *B(s)/A(s)* and *D(s)/C(s)* in Figure 10 are scalar.

$$Y(j\omega) = \frac{B}{A}U(j\omega) + W_2(j\omega) \tag{23}$$

$$U(j\omega) = \frac{D}{C}Y(j\omega) + W_1(j\omega) \tag{24}$$

then

$$Y(j\omega) = \frac{B}{A}(\frac{D}{C}Y(j\omega) + W_1(j\omega)) + W_2(j\omega) = \frac{B}{A}\frac{D}{C}Y(j\omega) + \frac{B}{A}W_1(j\omega) + W_2(j\omega), \quad \text{and}$$

$$Y(j\omega)(1 - \frac{B}{A}\frac{D}{C}) = \frac{B}{A}W_1(j\omega) + W_2(j\omega)$$

$$Y(j\omega) = \frac{\frac{B}{A}W_1(j\omega) + W_2(j\omega)}{(1 - \frac{B}{A}\frac{D}{C})} \tag{25}$$

$$U(j\omega) = \frac{\frac{B}{A}W_2(j\omega) + W_1(j\omega)}{(1 - \frac{B}{A}\frac{D}{C})} \tag{26}$$

for white independent noise w_1 and w_2.

The denominators of the quotients (25) and (26) are the *bias* of the transfer functions. Defining, $E(W_1^*W_1) = R$, $E(W_2^*W_2) = Q$ and $E(W_1^*W_2) = 0$, then

$$E(U^*Y) = \frac{(\frac{B}{A}R + \frac{D}{C}Q)}{\left(1 - \frac{B}{A}\frac{D}{C}\right)^2} = N \tag{27}$$

$$E(U^*U) = \frac{(\frac{D}{C}\frac{D}{C}Q + R)}{\left(1 - \frac{B}{A}\frac{D}{C}\right)^2} = M \tag{28}$$

$$E(Y^*Y) = \frac{(\frac{B}{A}\frac{B}{A}R + Q)}{\left(1 - \frac{B}{A}\frac{D}{C}\right)^2} = S \tag{29}$$

From this knowledge of cross correlations there is insufficient knowledge to find Q, R, $\frac{B}{A}$, $\frac{D}{C}$, since there are three equations (27) (28) (29) and four unknowns R, Q, $\frac{B}{A}$ and $\frac{D}{C}$.

If $\frac{D}{C}$ is known, the number of unknowns is reduced from (23)-(24) and W_1 can be found.

Now w_1 is an input and the transfer functions w_1 to Y and U can be calculated as in [29]

$$\frac{E(w_1^{*}Y)}{E(w_1^{*}U)} = \frac{\frac{B}{A}R}{R} = \frac{B}{A}$$

If the feedforward dynamics composed by $\frac{B}{A}$ and $\frac{D}{C}$ with $\frac{D}{C}$ scalar are known, the same process yields the transfer function. Once the high frequency feedback gain is known, then the loop can be resolved and the transfer function for the load derived.

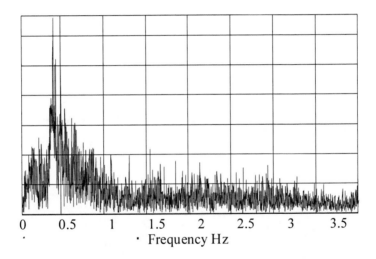

Figure 15. Frequency changes affecting load P

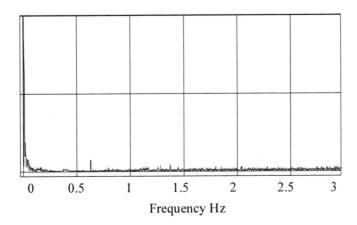

Figure 16. Load P changes affecting frequency

3.1.3. Industrial example of frequency to power relations

An example of the process applied to an industrial load in Tasmania is seen in Figure 15 and Figure 16 where U is measured bus frequency and Y is measured load real power P. The feedback term in Figure 16 shows a governor response characteristic of the frequency response of Tasmania to load changes. There are no electromechanical modes excited by this load, even the expected 1Hz North South oscillation mode is absent since the load is at the node of this mode.

There is a significant difference between the two figures. If the process of separation of feedforward and feedback were imperfect the poles of both would appear in the inferred decomposition as could be seen if a simple direct transfer function from f to P was indentified.

3.1.4. Industrial example of voltage magnitude to power relations

If the system functions $\dfrac{B}{A}$ and $\dfrac{D}{C}$ were not low pass and the load noise was normally distributed, then the previous analysis would indicate that decomposition would not be possible. There is one saving factor for real system however which has been tested across four different sites which indicates that the noise shows a higher probability of large excursions than expected from a Normal Distribution. This is shown in Figure 17 where a straight line fit would indicate a Normal distribution. This shows that real power systems do not follow the law of large numbers where the sum of many distributions becomes normal if they are of similar size. Here the existence of a few large loads means that the variations are not Normal distributions. This allows the possibility of having distinct customer load changes in the feeder separable from isolated events originating from the rest of the system.

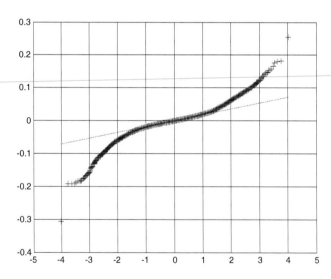

Figure 17. QQ plot showing deviation from Normal distribution.

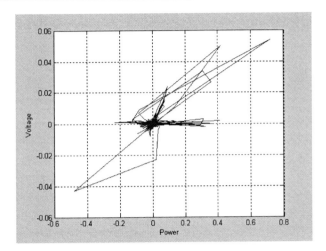

Figure 18. Plot of change in V against change in P.

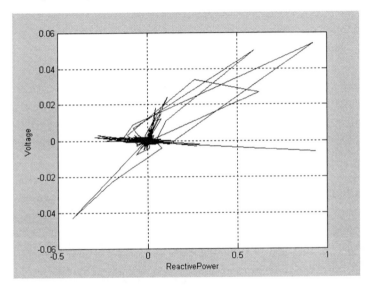

Figure 19. Plot of change in V against change in Q.

Plots in Figure 18 and Figure 19 show a set of excursions of the load P and Q changes respectively with a slight negative slope in the voltage change. Another set of excursions with a distinct positive slope can be found. The interpretation of this behaviour is that the slight negative slope is the voltage drop caused by feeder load steps P and Q causing small drops in the voltage at the substation. Positive slope is interpreted as an external voltage changes causing a rise in load power. The higher negative slope in Figure 19 is reassuring in that the supply impedance is expected to be largely reactive. From these figures the direct feedthough gain in $\dfrac{D}{C}$ of Figure 10 can be identified. The identification of separate feedforward and feedback components now becomes possible because of these large impulse load variations which are distinctly within the feeder.

3.2. Improving the Small Signal Model: F-V Relationship

The approach presented so far considers there is no relation between voltage and frequency. But this is not the real case.

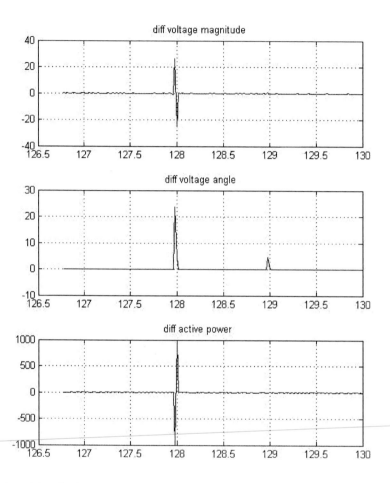

Figure 20. Voltage Magnitude, Frequency and Power for two major real events

Analysis of data collected from a major event and a voltage spike revealed a relationship between frequency and voltage magnitude. A graphic of the data is shown below in Figure 20. As explained before the frequency variations are related to the phase angle of the voltage. The relation between frequency inducing changes in voltage is clear in this figure and that means there is a strong relationship between frequency and voltage magnitude.

This relationship must be taken into account to obtain a more accurate model of voltage and frequency changes inducing changes in active and reactive power.

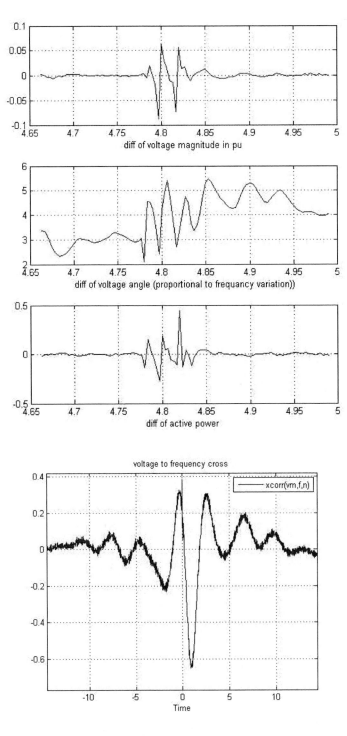

Figure 21. Cross-correlation between V and f.

The cross-correlation between frequency and power is shown in Figure 21 for real data which didn't present a major disturbance. This figure shows a non-causal relationship between frequency and voltage.

The general case where the voltage magnitude affects frequency and power is shown in Figure 22 where the model is formulated. Figure 23 shows the correspondence of this model with the power system case.

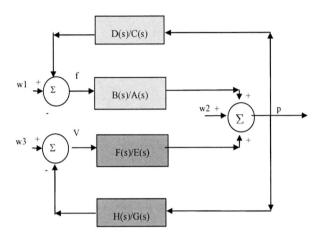

Figure 22. Two inputs to One output transfer function.

Coloured noise will appear as result of the voltage-frequency relationship if the simple case of closed loop is used to describe the frequency to real power (f-P) relationship. A possible realization of this influence is shown in Figure 24.

Considering the case with two inputs, U and V with associated noises w_1 and w_3, and one output, Y with associated noise w_2.

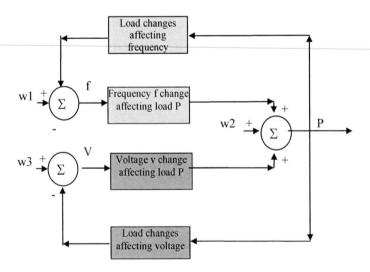

Figure 23. Power system case.

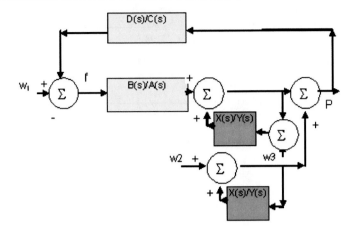

Figure 24. Realization of the influence of voltage on frequency as coloured noises.

The relations between inputs and output are

$$Y(s) = \frac{B(s)}{A(s)} U(s) + \frac{F(s)}{E(s)} V(s) + W_2(s)$$
(30)

$$V(s) = \frac{H(s)}{G(s)} Y(s) + W_3(s)$$
(31)

$$U(s) = \frac{D(s)}{C(s)} Y(s) + W_1(s)$$
(32)

Substituting (30) into (31)-(32), the variables as function of the noises are

$$Y(s) = \left(\frac{B(s)}{A(s)} W_1(s) + \frac{F(s)}{E(s)} W_3(s) + W_2(s) \right) \Big/ \left(1 - \frac{B(s)}{A(s)} \frac{D(s)}{C(s)} - \frac{F(s)}{E(s)} \frac{H(s)}{G(s)} \right)$$
(33)

$$V(s) = \left(\frac{H(s)}{G(s)} \frac{B(s)}{A(s)} W_1(s) + \frac{H(s)}{G(s)} W_2(s) + W_3(s) \left(1 - \frac{B(s)}{A(s)} \frac{D(s)}{C(s)} \right) \right) \Big/ \left(1 - \frac{B(s)}{A(s)} \frac{D(s)}{C(s)} - \frac{F(s)}{E(s)} \frac{H(s)}{G(s)} \right)$$
(34)

$$U(s) = \left(\left(1 - \frac{F(s)}{E(s)} \frac{H(s)}{G(s)} \right) W_1(s) + \frac{D(s)}{C(s)} \frac{F(s)}{E(s)} W_3(s) + \frac{D(s)}{C(s)} W_2(s) \right) \Big/ \left(1 - \frac{B(s)}{A(s)} \frac{D(s)}{C(s)} - \frac{F(s)}{E(s)} \frac{H(s)}{G(s)} \right)$$
(35)

Equations (33)-(35) show bias terms associated with the other input appearing in these equations.

Applying cross-correlations as in equations (33)-(35) and considering the noises w_1, w_2 and w_3 are not correlated; the transfer functions can be obtained as

$$\Phi_{YU}(j\omega) = \frac{D(j\omega)}{C(j\omega)} = \left(\frac{(tf(w_2 : u))}{(tf(w_2 : y))}\right) \tag{36}$$

$$\Phi_{YV}(j\omega) = \frac{H(j\omega)}{G(j\omega)} = \left(\frac{(tf(w_2 : v))}{(tf(w_2 : y))}\right) \tag{37}$$

where $tf(w_2{:}y)$ represents the fast Fourier transformation (*fft*) calculated from real measurements from the noise w_2 to variable y. This can be calculated using Matlab algorithms. The other relationships involving *fft* have the same meaning regarding the variables. Note that more than one approach can be used to calculate feedback terms.

One way the transfer function for the feedforward terms can be obtained is

$$\Phi_{UY}(j\omega) = \frac{B(j\omega)}{A(j\omega)} = \left(\frac{(tf(w_1 : y)\, tf(w_3 : v) - tf(w_3 : y)\, tf(w_1 : v))}{(tf(w_1 : u)\, tf(w_3 : v) - tf(w_1 : v)\, tf(w_3 : u))}\right) \tag{39}$$

$$\Phi_{VY}(j\omega) = \frac{F(j\omega)}{E(j\omega)} = \left(\frac{(tf(w_3 : y)\, tf(w_1 : u) - tf(w_1 : y)\, tf(w_3 : u))}{(tf(w_1 : u)\, tf(w_3 : v) - tf(w_1 : v)\, tf(w_3 : u))}\right) \tag{40}$$

3.2.1. Numerical example

The proposed approach was tested in Matlab with the following transfer functions with sampling time 0.2 sec and transfer functions from input u and v to output y:

$$\frac{Y}{U} = \left(\frac{-0.8817\,z + 0.9178}{z^2 - 1783z + 08187}\right) \quad \text{and} \quad \frac{Y}{V} = \left(\frac{0.1801\,z - 0.1801}{z^2 - 1783z + 08187}\right)$$

$$\frac{U}{Y} = \left(\frac{0.2214}{z - 0.6110}\right) \quad \text{and} \quad \frac{V}{Y} = \left(\frac{0.2459}{z - 0.7460}\right)$$

The feedback terms are

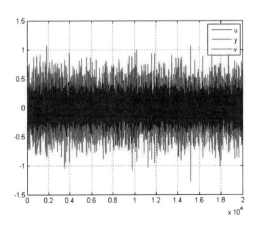

Figure 25. Variables in time domain.

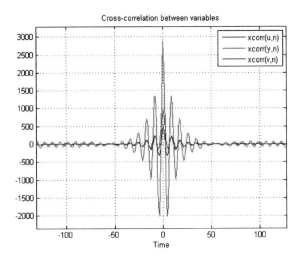

Figure 26. Cross-correlation between the variables.

Figure 25 shows a realization of the variables in time domain and Figure 26 shows the cross-correlation between the variables. The cross-correlation shows a non-causal system.

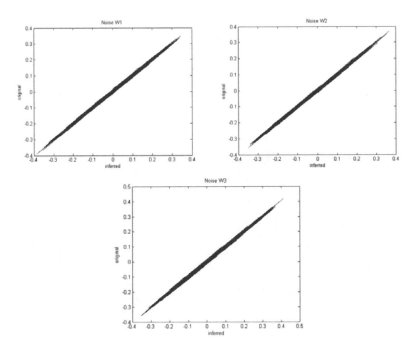

Figure 27. Inferred and Original noises.

The proposed method is based on the assumption that the noises can be extracted successfully from the variables. For this theoretical case, the following figures show the relation between the original and inferred noises for this realization.

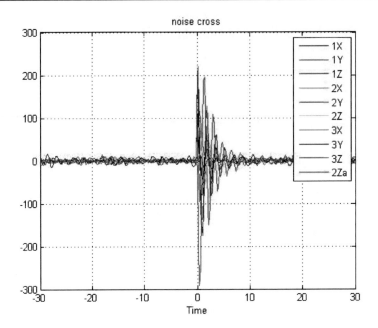

Figure 28. Cross-correlation between noises and variables.

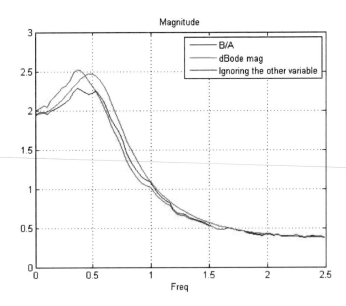

Figure 29. B/A feedforward term.

Figure 28 shows the cross-correlation between the variables and the inferred noise, showing there is causal relationship between the extracted noises and the other variables.

Feedforward calculated and actual terms are shown in Figure 29 and Figure 30 . Calculated terms considering the influence of the other variable are plotted in blue and terms that do not consider the other variable in red. In green the terms obtained with dBode algorithms of Matlab show the actual transfer function. There is an error in the magnitude of the lower magnitude function F/E.

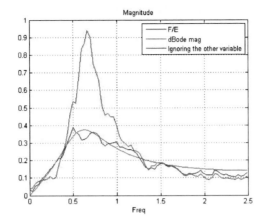

Figure 30. F/E feedforward term

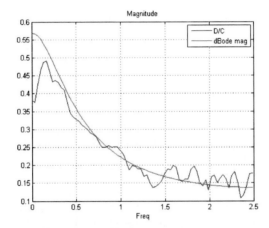

Figure 31. D/C feedback term

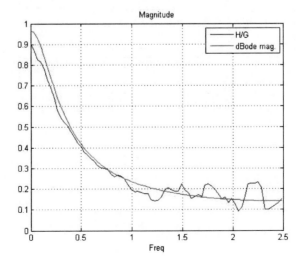

Figure 32. H/G feedback term.

Figure 31 and Figure 32 show the D/C and H/G feedback terms.

Figures 33 to 36 show the phase of the calculated and actual feedforward and feedback terms.

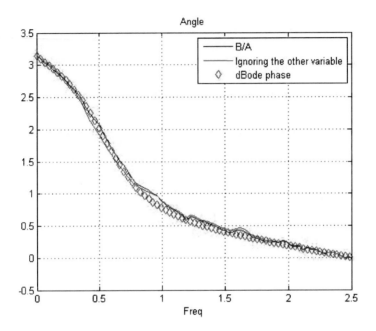

Figure 33. Phase of the B/A term.

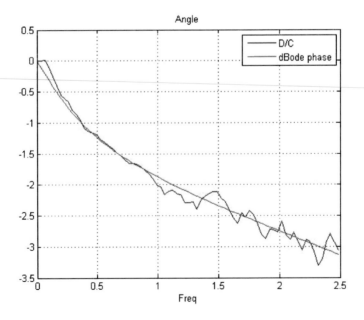

Figure 34. Phase of the D/C term

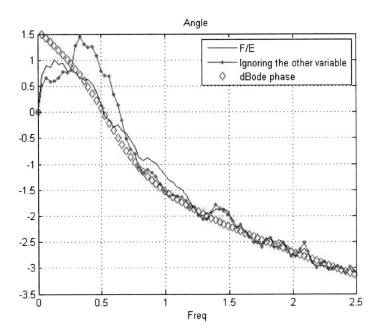

Figure 35. Phase of the F/E term

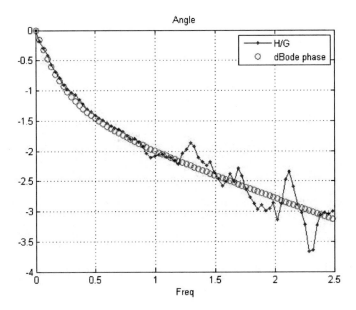

Figure 36. Phase of the H/G term

From the previous figures it can be concluded that not considering the bias introduced by the other variable induces an error, and this error depends on the transfer function of the other variable. Figure 37 shows the bias introduced by the other variable. For the B/A case the magnitude of the bias remains around one for all the frequencies, but in the case of F/E the bias magnitude has a strong influence on the value of the transfer function.

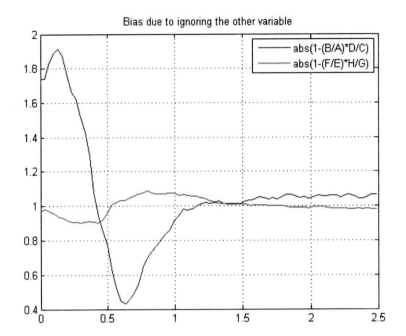

Figure 37. Bias introduced by the other variable

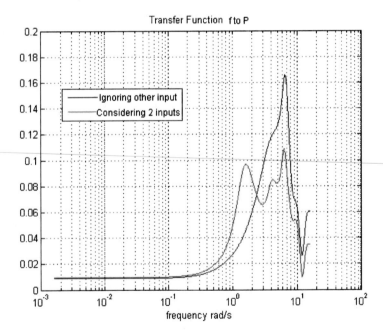

Figure 38. Transfer function f to P

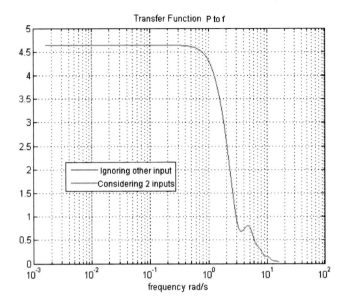

Figure 39. Transfer function P to f

3.2.2. Results when applied to real Frequency and Voltage measurements

Transfer functions obtained from real measurements are shown in the next figures, calculated with the frequency domain approach and using a Tuckey window (16, 0.5), considering the influence of the other variable and not. The transfer function B/A is the frequency f to real power P transfer function, D/C is the P to f transfer function, F/E is the transfer function from voltage magnitude V to P and H/G is the transfer function from P to V. The transfer functions of feedback terms are similar in both cases.

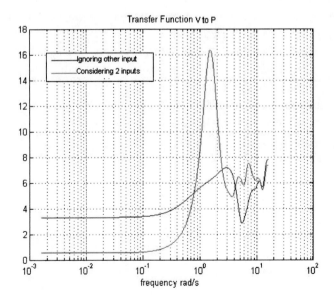

Figure 40. Transfer function V to P.

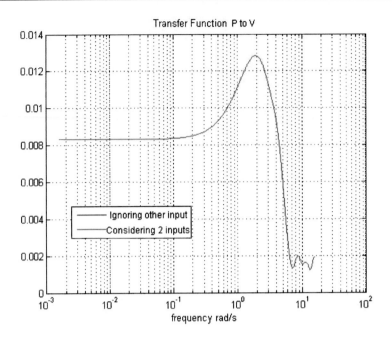

Figure 41. Transfer function P to V.

4. IDENTIFICATION OF MOTOR LOADS

The identification of motors load is can be performed using the f to P transfer function. The variation in power caused by variations in frequency are directly associated with induction motor load. The basic idea is to create groups of motors related by nominal power and inertia. Fitting the f-P transfer functions of individual motors into the f-P transfer function of the feeder is one way to characterize the load.

The 5th, 3rd and 1st order models described in section 2 for induction motor simulation have their transfer functions from frequency to active power plotted in Figure 42.

As can it be seen the 5th and 3rd models show poles that represent the dynamics from rotor and stator transients. But in Figure 42 it is clear that the magnitude of the peak of the f to P transfer function is the same for 1st, 3rd and 5th order model.

The transfer function between frequency changes and active power for the 1st order model can be expressed as

$$\frac{\Delta P_e}{\Delta f} = \frac{V^2}{R_r \omega_s} \frac{s + B/2H}{s + \frac{B}{2H} + \frac{V^2}{R_r \omega_s}} \tag{41}$$

where B is the load of the induction motor.

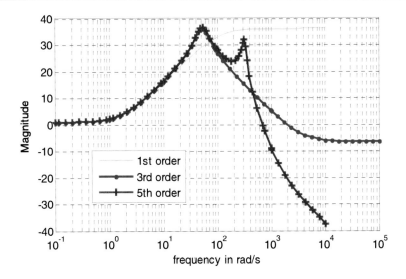

Figure 42. Magnitude of induction motor frequency to Power transfer function

The high frequency gain of the first order transfer function is expressed as $V^2/(R_r\omega_s)$. If the slip is known then the value of the peak will be proportional to the power of the motor [36].

Since in all the examined cases the zeros of the system are sufficiently far from the peak, the magnitude of the real portion of the peak closely determines the area under the curve of real(f-P) when plotted against log (ω) [37].

In [37] a set of induction motor parameters from 4kW to 630kW taken from [18]. is evaluated on the p.u. base of each machine. The real component of the frequency to power transfer function becomes as in Figure 43 where '1' is the lowest power machine and '10' is the highest. For each of these machine the peak of the plot is proportional to $V^2/(R_r\omega_s)$. The lower inertia machines have a higher rated slip and a higher frequency of the peak [38].

The total area under the real part of the f-P curve is a close indication of the total power of the component associated with induction motors. The real component of the transfer function is used to avoid the effect of the phase when different motors are added.

The real part of the transfer function from frequency to real power allows identifying the motors present in the load [38]. This can be performed fitting the real part of the f-P transfer function of small, medium and large motors into the real part of the composite transfer function. The percentage contribution of each group of motor can be found by using the least squares algorithm.

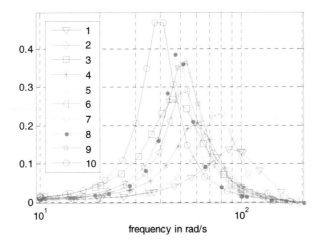

Figure 43. Real component of motor frequency to Power transfer function from [38]

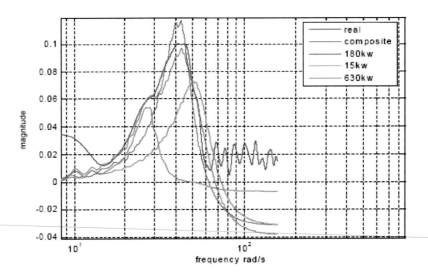

Figure 44. three motor templates and the composite from [38].

Changes in the motors' inertia shift to right and left the transfer function in the frequency domain. This fact is used to identify the inertias of the motors. Testing different inertias associated to different motors' nominal powers is the process to fit the composite models to real. The motors' real magnitude of the transfer function values are selected as template motors and are used as the input of the least square input X and the value of the real data transfer function as an output Y. Least squares identification algorithm is used to decompose the measured composite motor responses. To avoid negative components the following optimization problem is proposed,

$$\min \quad (X\theta - Y) \qquad\qquad (42)$$

$$st \quad \theta > 0$$

Here, Y is the real magnitude of aggregated real power to frequency change transfer function. θ is the percentage contribution of each motor and X is the real magnitude of individual motor's frequency change to real power change transfer function. To avoid the possibility of having values of percentages of motors with negative value a lower limit to the variables θ was imposed. The percentage contributions of each machine are estimated from this problem.

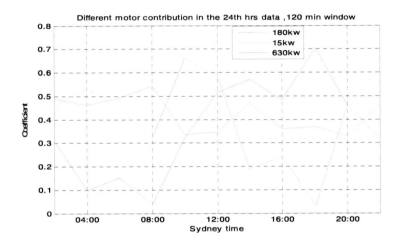

Figure 45. Motor composition of real data – 120 min window.

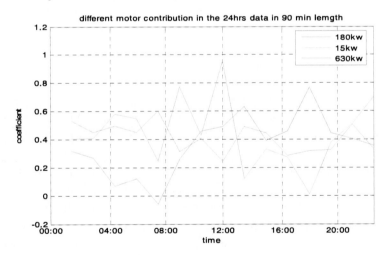

Figure 46. Motor composition of real data – 90 min window.

The same three template motors with modelled data is used to determine a composition of different motors in 24 hours of Sydney West data in [38]. The percentage contributions of the three template motors in 24hrs data are shown in Figure 45 and Figure 46 below. Two different window lengths of 90 and 120 minutes are used to calculate the composition. The reasons behind using different window lengths are that each window length can retain the real load changing phenomena in a feeder.

The variation of the small motor group is quite substantial in the 120 mins data window lengths in Figure 45. From Figure 45 it can say that the small types of appliances are switching on and off in 24hrs time frequently whereas the large motors 630kW is switching on from approx 8:00 am until 6:00 pm and then it starts to be switched off. The large type of motors is used in industry and it is evident that the big motor is turned on during daytime and turned off in the evening.

5. LARGE DISTURBANCE

The response of induction motors to large disturbances is conditioned by the power system structure, the amount of motors present at the time of the disturbance, and the nature of any load protection units.

The appropriate model needs to be considered when motors are simulated. The single cage model is not accurate to calculate the starting torque and current, but it presents a close approximation near synchronous speed. As shown, the maximum torque is dependent of the square of the input voltage and the lower torque at low speed means the motor has a great expectation to stall under a deceleration caused by voltage sag. If the same voltage dip is applied to the double cage model, it is probable that it could ride through [20]. When power system loads are modelled, the set of motors are represented by a number of aggregate motors that have to simulate thousands of motors. An adequate combination of both types of motor is used in references [24, 25]. Those references use the fact that an adequate amount of motors need to be stalled to represent appropriately the voltage recovery delay [24, 25]. Numerical stability of the algorithm is a clear objective in [25].

The next simple example shows an approximation to the real behaviour. The example was created in PSCAD 4.2 and consists in 5 identical induction motors 11KW/0.4KV with the same inertia and two 90 KW/0.4KV connected to the same low voltage line. The connection of the transformer is shown in the figure and the parameters of the *pi* sections representing the lines were extracted from design data of a real industrial facility. Induction motor parameters used in this simulation where taken from [16]. The diagram is shown in Figure 47.

A three phase fault is applied at t=20 sec for 0.5 sec, allowing the motors to be started and working properly. The fault is applied to another branch of the grid, and it produces voltage sag on the feeder where the induction motors are connected. The voltage at the sag is approximately 40% of the pre-fault voltage.

Figure 48 shows the active and reactive power of motors 1 and 5, the voltage at their terminals and the variation of speed during the process. As it can be seen the presence of the transformer impedance is the only impedance between motor 1 and the transformer 250KVA 2% impedance, so the voltage at the transformer terminals is the same. The presence of the systems' impedance delays the recovery of the voltage at the terminals of motor 1.

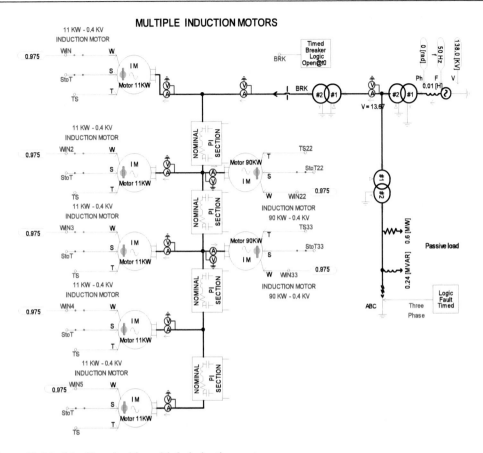

Figure 47. Model of Load with multiple induction motors.

Motor 5, placed at the end of the line, is affected in a more dramatic way. The voltage sag is deeper, driving the motor to total deceleration and it is affected by the reacceleration of the others motors when the voltage recovers. In this case motor 5 tries to reaccelerate from zero speed, and since the voltage at its terminal is affected by the recovery of motors 1-4 and 6-7, the amounts of active and reactive power demanded are greater than motor 1 and for a longer time. Usually large nominal power motors have the possibility of avoiding the reacceleration and since a ride-through of the voltage sag is not beneficial for the motor, adequate protection settings will disconnect these motors from the grid [41]. Another consideration is that motors reaccelerating under unfavourable conditions, such as under-voltage recovery, will have higher heat dissipation and if the overload relay is properly set, the protection will open before the motor reaches nominal speed. Note that the time between the clearing of the fault and the reacceleration of the motors is around 4 seconds. The simulated motors are double cage type. Instability was observed on motor 5 at the speed recovery for different values of voltages.

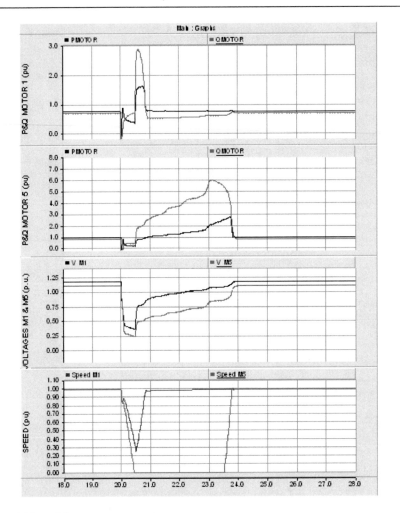

Figure 48. Variables of the motors.

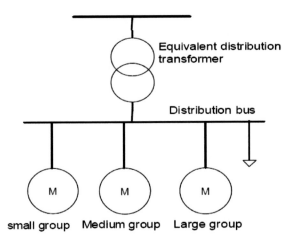

Figure 49. Equivalent model for a major disturbance

5.1. Simulating Major Disturbances

The curve of Figure 48 can be replicated by motors connected to the bus as in the Figure 49.

To replicate the major disturbances at least the following points must be assessed

1) Nominal power of the motors.
2) Inertia of the motors
3) Numerical stability of the model

The information available, which can be gathered from small signal identification, is just nominal power and inertia of the motors presented in the pre-fault state. Even if those parameters were known with high accuracy, to replicate the disturbance involves numerical problems that need to be understood. Some of the identified issues involving large scale simulation are:

1) The double cage induction motor model is more robust and has less stability problems that the single cage model. However, cases of instability have been reported in [39].
2) The grid impedance between the bus where the measurement can be taken and the motor 5 is the main source of delay of recovery. This is a similar situation than presented in [24, 25], where the grid effect is accentuated by stalled motors.
3) The magnitude and duration of the voltage dip can be used to identify stalled motors.

The severity Index and the Tripping Index, presented in the next sections, are formulated to characterize the load associated with induction motors.

5.2. Severity Index

The aim of a severity index is to predict what fraction of motors at an aggregate bus will stall for a given voltage dip and length. This section considers first the characterization of single motor stalling and proceeds to the application to aggregate motors.

Since the induction motor torque is dependent of the square of the voltage, a relation between the non-delivered energy in the voltage dip and the loss of speed of induction motors can be expressed as a function of the voltage dip characteristics, such as voltage dip and duration.

The characterization of the fraction of motors in the pre-disturbance state as a function of their nominal power allow the use of the severity index, taken into account that the "prone to stall" motors are commonly the lowest inertia motors.

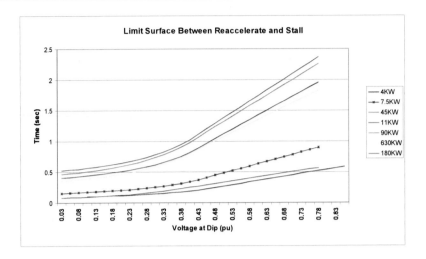

Figure 50. Limit Surfaces between reaccelerated and stalled IM

The severity index can be calculated using simulation over different types of motors, since the size of the motors present in the load can be identified.

1) Obtain via simulation the points of the curve Ω that defines the border of the regions between stalled and surviving motors.

2) Calculate the coefficients K1 and K2 that approximate the curve. Voltage at the pre-fault state is considered constant in the simulations.

$$\Omega(\Delta t) = K_1 - K_2 \left(V_{no\,min\,al}^2 - V_{fault}^2(\Delta t) \right) \qquad (43)$$

Where Δt is the time duration of the disturbance, $V_{pre-fault}$ is the pre-fault voltage of the bus, V_{fault} is the voltage during the fault and K are the factors that adjusts the generic curve to each motor.

The coefficients K are calculated using the least square approximation, for each motor. They provide an index that allows reproducing the limit curve between stalled and reaccelerating states of the motor to a generic curve. They values grows when the motor is less susceptible to stall.

The coefficients K are the result of the problem:

$$\min \sum \left(K_1 - K_2 \left(V_{no\,min\,al}^2 - V_{fault}^2(\Delta t) \right) - X_{data} \right)^2 \qquad (44)$$

In the following simulations the single cage was used. The voltage sags considered are three-phase voltage dips. Other types of voltage sags were not considered at this stage of the research. Unbalanced voltage sags can be treated as in [46, 47].

Figure 50 shows the limit surface between stalled and reaccelerated motors for the some of the motor data published in [18]. As can it be seen, there is not a clear correspondence between the nominal power and the ability of an induction motor to ride through a voltage dip. This can be explained looking at

Table 3 where the relation between initial torque and nominal torque for NEMA motors is shown. In that table large motors have a lower relation between nominal torque and start torque. Figure 50 is also coherent with simulations carried out by [39] where 11kW and 630 kW motors were more susceptible to stall under a voltage dip that the 180KW motor. In that reference it is shown the effect of the inertia of the motor on the single cage model under the same voltage dip. Modifying the inertia of the motor plus mechanical load means modifying the curves. Table 4 shows the effect of increasing the inertia 10 times for a voltage dip of 0.1pu.

Table 4. Time to stall the motor for a voltage dip=0.1 pu considering 10 times the original inertia

Induction Motor	4KW	7.5KW	15KW	22KW	45kw	180KW	630KW
Time (sec)	1.6036	2.1377	1.1337	2.7603	2.516	4.5989	6.1059

Table 5. Severity Index Values

Induction Motor	4KW	7.5KW	45KW	22KW	630KW	180KW
Constant term K_1	0.8395	1.1989	0.8611	1.1217	1.7347	2.1518
Linear term K_2	0.7575	1.0783	0.7933	1.008	1.5374	1.8584

From Table 4, it is interesting to observe the behaviour of the 630kW motor. For the same voltage level at the dip this motor would be stalled if the sag last for more than 0.23 sec. When the inertia is increased 10 times, the motor would be stalled for the same voltage level at the sag if the sag last for more than 6.1 sec. This example points out the importance of considering the type of mechanical load attached to the motor and different inertia values in the simulations. Compressors' inertia can be considered as represented approximately by the same inertia of the motor, but this can not be taken as a rule especially for large motors. For the simulation carried out in Figure 50 the values of the Severity Index are shown in Table 5.

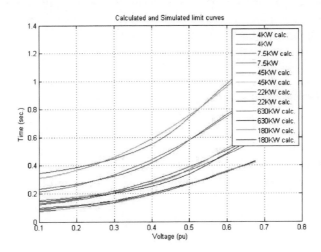

Figure 51. Simulated and Calculated curves

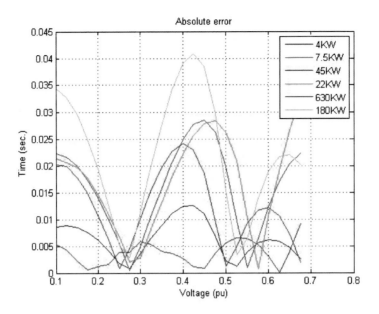

Figure 52. Absolute error of using the approximated curves

The match between simulated motors and quadratic curves formulated in Equation (44) are presented in Figure 51. The absolute error of the approximation is presented on Figure 52, where the error for the 180KW motor for a voltage at the dip of 0.2 pu is 0.02 sec, or the 5% of the time needed to stall the motor. Figure 53 shows the relative error in time if the approximation of the curve is used instead of the simulated values.

Since the parameters of the induction motors can be obtained using the nominal power, it can be considered that the severity index for each type of motor is unique for each inertia motor-load value. Once the severity index is known the state of the motor can be defined using the voltage dip as input. Note that if the severity index is known just one point is necessary to determine the limit between stalled and reaccelerated motors.

The severity index for one motor is a powerful tool to determine the state of a motor using the voltage dip as input. It allows determining the state of the any motor to be used in the simplified model presented in Figure 49.

5.3. Tripping Index

The tripping index can be defined considering that a common set of protections of induction motors includes a contactor plus a protection to overload and short-circuits. Protection to overload and short-circuits are based on inverse time and maximum current. Thermo-magnetic protections follow this rule.

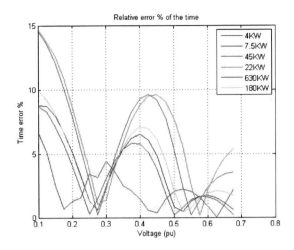

Figure 53. Relative error of using the approximate curves.

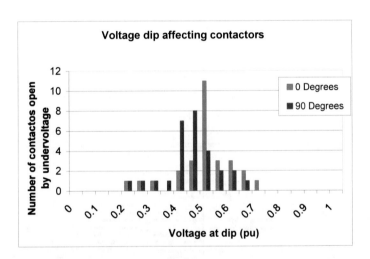

Figure 54. Acquired distribution of drop-out voltage for 28 contactors at two point-on-wave values [45].

Contactor-relay sets were tested in [45] according to the European norm IEC 60947-4-1. This standard for electromagnetic contactors establish that contactors shall close satisfactorily at any value between 85% and 110% of their rated control supply voltage and shall drop out and open fully between 75 % and 20 % of U for AC, 75 % and 10 % for DC. These limits refer to the steady state conditions, since no time limits are given.

The point-on-wave of initiation strongly affects the contactor performance. This is due to the energy stored in the magnetic circuit as a remaining (dc) flux. The stronger the flux is at the moment of sag initiation the better the contactor rides through the event. Further, the deeper the sag, the more significant the decaying dc component of the magnetic field is. At the moment of voltage maximum (90°) the momentary current and remaining dc flux are close to zero. This leads to the conclusion that contactors are most sensitive to sags initiated at 90° and least sensitive to sags initiated at 0°. In practice the maximum flux is achieved at the voltage zero-crossing (0°) [45]. From a utility point of view the range of voltage sags due to disturbances on the grid is between 50 to 500 ms, which is coherent with the protection settings. Within this range the point-on-wave does not play a mayor role neither does the duration. The magnitude remains practically the only interesting measure. The distribution chart of this steady-state voltage limit for two point-on-wave angle values (0° and 90°) in Figure 54 shows the performance of the 28 tested contactors by the same author. Most tolerate 50 %, a commonly proposed level. Figure 54 also clearly indicates that there is a wide range of selectivity between the contactors.

The results from [45] allow understanding the instantaneous tripping of induction motor loads under voltage sags. But the load used by [45] in the tests was discharge lamps which are not representative of the induction motor load. As shown the non-tripped motors would try to reaccelerate and the successful reacceleration of the motor depends on the thermo-magnetic protection setting and the electrical distance to the source, if it is not stalled.

The inverse time of the curve can be represented by the following equation,

$$t(I) = \frac{A}{\left(\dfrac{I}{I_p} - 1\right)}$$

(45)

Where t is the time, I is the current to the motor, I_p is the pickup current and A is a constant that represent the relation between the drag magnet and the spring torque [40]. This equation means currents above the pickup current would trip the protection at time t.

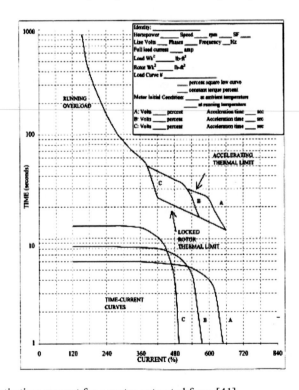

Figure 55. characteristic time-current for a motor extracted from [41]

Figure 55 shows the characteristic time-current for a motor extracted from [41].

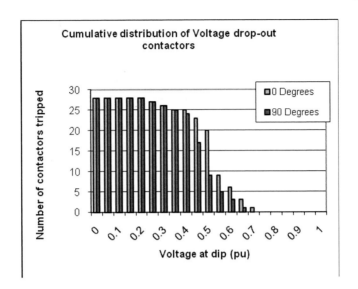

Figure 56. Cumulative of the probability distribution for 28 contactors with two point-on-wave values

As shown before, after a voltage dip, motors far away from the source would try to reaccelerate but their high current drain causes a delay. When the motor is electrically close to the source its current at the voltage recovery can be considered as equivalent to the start current. The low voltage experimented at the terminals of induction motors long from the source depress the currents in the post-sag scenario and the result of this process is that they would trip in a longer time. The restart of large numbers of motors could stall many of the motors, especially if they are small inertia motors. In the last case, currents similar to the start current would be reached as the voltage recovers which means the recovery time would be extended.

Curves similar to the presented at the severity index can be calculated for each motor. The main problem is how to disclose the physical location of the induction motors, which is how far away electrically are they from the source.

If data from the grid is available, accurate simulations can be performed and the total energy demanded by the reacceleration of the motor or the stalls of the motor can be calculated. Motor 5 in Figures 47 and Figure 48 can not survive and will trip. The specific energy, i^2t, presented in [43] allows predicting the opening of the thermo-magnetic protection. A similar process is presented in reference [44] where a curve reaccelerating-tripping is obtained for an induction motor.

The Tripping index can be formulated as the product between the accumulative of the probability curve of the contactors' trip shown in Figure 56 with the over-current protection. Simulation of the motor behaviour must reflect the interactions between motors in the grid. Also, the probability of under-voltage protection present in large motors need to be assessed.

The combination of the Severity Index and Tripping Index allows defining the equivalent motor loads that need to be considered tripped or stalled to simulate accurately the response of the motor load to voltage dips.

Next table shows the timing to trip for the 7.5 KW induction motor presented in Figure 50. The test was performed considering a stiff bus. In bold appear the states of the motor where it is stalled. As can be seen the inverse time protection did not trip when the duration of the dip is less than 0.3 secs. There are states where the relay trip because of the current demanded by the motor and others where the motor is stalled. Since the motor was simulated against a stiff bus the delay of voltage recover that appear at Figure 48 is not considered.

From the last Table it can be seen that considering just the relay is not accurate enough because the contactors would trip under low voltage as shown before.

The following algorithm can be used to check the accuracy of the simulation in real disturbances by obtaining the percentages of motors reaccelerating, stalling or tripping in the pre-fault state and the post-fault state using real measurements and the algorithms presented in the small signal model to disclose the component of induction motors.

Table 6. Times to trip a 7.5 KW induction motor after a voltage dip.

Voltage at dip	0.1 sec.	0.2 sec.	0.3 sec.	0.4 sec.	0.5 sec.	0.6 sec.	0.7 sec.	0.8 sec.	0.9 sec.	1 sec.
0.1	No trip	No trip	No trip	0.25	stalled	stalled	stalled	stalled	stalled	Stalled
0.2	No trip	No trip	No trip	0.25	0.24	stalled	stalled	stalled	stalled	Stalled
0.3	No trip	No trip	No trip	No trip	0.26	0.25	stalled	stalled	stalled	Stalled
0.4	No trip	No trip	No trip	No trip	No trip	No trip	0.3	0.3	0.3	Stalled
0.5	No trip	No trip	No trip	No trip	No trip	No trip	No trip	No trip	No trip	0.3
0.6	No trip	No trip	No trip	No trip	No trip	No trip	No trip	No trip	No trip	No trip
0.7	No trip	No trip	No trip	No trip	No trip	No trip	No trip	No trip	No trip	No trip
0.8	No trip	No trip	No trip	No trip	No trip	No trip	No trip	No trip	No trip	No trip
0.9	No trip	No trip	No trip	No trip	No trip	No trip	No trip	No trip	No trip	No trip
1	No trip	No trip	No trip	No trip	No trip	No trip	No trip	No trip	No trip	No trip

5.4. Decomposing the Response of Motors in Stalled, Reaccelerated and Tripped

The active and reactive power associated with induction motors (P_B^{total} and Q_B^{total}) in the pre-disturbance state can be defined as

$$P_{lB} + P_{sB} + P_{tB} = P_B^{total}$$

(46)

$$Q_{lB} + Q_{sB} + Q_{tB} = Q_B^{total}$$

(47)

where P_{lB}, P_{sB}, P_{tB} and Q_{lB}, Q_{sB}, Q_{tB} are the components of active and reactive power related to reaccelerated, stalled and tripped motors. After the disturbance the component corresponding to the tripped motor is not present.

$$P_{lA} + P_{sA} = P_A^{total}$$

(48)

$$Q_{lA} + Q_{sA} = Q_A{}^{total} \qquad (49)$$

where P_{lA}, P_{sA} and Q_{lA}, Q_{sA} are the components associated with active and reactive power for reaccelerated and stalled motors after the disturbance, and $P_A{}^{total}$ and $Q_A{}^{total}$ are the total measured active and reactive powers after the disturbance.

Since the number of equations is greater than the number of variables, the following considerations need to be addressed:

a)   In a stalled motor the amount of reactive power will be higher than the active power component making it a reasonable assumption that the active power component is zero. The reactive power demanded by the stalled motors can be considered greater than the reaccelerated motors.

b)   Active and reactive power of reaccelerated motors are the same as before the disturbance, and reactive power of reaccelerated motors can be approximate to zero if compared with stalled motors.

Taken into account the assumptions a) and b), the system of equations (46)-(49) can be rewritten as

$$P_{lB} + P_{sB} + P_{tB} = P_B{}^{total} \qquad (50)$$

$$Q_{lB} + Q_{sB} + Q_{tB} = Q_B{}^{total} \qquad (51)$$

$$P_{lB} = P_A{}^{total} \qquad (52)$$

$$\beta Q_{sB} = Q_A{}^{total} \qquad (53)$$

Fixing a value for β, the reactive power associated with stalled motors, can be calculated as $Q_{sB} = Q_A{}^{total} / \beta$, and active power for reaccelerated motors is $P_{lB} = P_A{}^{total}$. Since $\cos(\varphi)$ of the whole load is known before the disturbance, the active power demanded by stalled motors and the reactive power of reaccelerated motors can be calculated for the pre- disturbance period. Finally, the active and reactive power component associated with the tripped motors can be calculated as $P_{tB} = P_B{}^{total} - P_{lB} - P_{sB}$ and $Q_{tB} = Q_B{}^{total} - Q_{lB} - Q_{sB}$.

The calculated values with this approach presented an average error greater than 15%, associated with the considerations.

The components associated to the post-disturbance state are recalculated using an optimization scheme, which can be described as.

$$\text{Min} \left(A^T A \right)$$

where

$$A = \begin{bmatrix} P_A^{total} - P_{lA} - \left(\dfrac{Q_{sA}}{\tan(\phi)_{stall}} \right) \\ Q_A^{total} - Q_{sA} - \left(P_{lA} * \tan(\phi)_{bef} \right) \end{bmatrix} \qquad (54)$$

This unconstrained optimization problem has as variables P_{lA} and Q_{sA}, and $\tan(\varphi)_{stall}$ and $\tan(\varphi)_{bef}$ are the typical tangent of a stalled motor and the tangent of the load before the disturbance.

The solution for this unconstrained problem is considering the gradient of $\left(A^T A \right)$ equal zero, that is,

$$\nabla_{P_{lA},Q_{lA}} \left(A^T A \right) = 0$$

where

$$\nabla_{P_{lA},Q_{lA}} \left(A^T A \right) = \begin{bmatrix} \dfrac{\partial A^T A}{\partial P_{lA}} \\ \dfrac{\partial A^T A}{\partial Q_{sA}} \end{bmatrix} = \begin{bmatrix} 2 * A^T * \begin{bmatrix} -1 \\ -\tan(\varphi)_{bef} \end{bmatrix} \\ 2 * A^T * \begin{bmatrix} -\dfrac{1}{\tan(\varphi)_{stall}} \\ -1 \end{bmatrix} \end{bmatrix} \qquad (55)$$

Equation (30) is solved via Newton Method. At each iteration the following linear system needs to be solved:

$$W \left(A^T A \right) * \begin{bmatrix} \Delta P_{lA} \\ \Delta Q_{sA} \end{bmatrix} = \nabla_{P_{lA},Q_{lA}} \left(A^T A \right) \qquad (56)$$

where

$$W = \begin{bmatrix} \dfrac{\partial^2 \left(A^T A \right)}{\partial^2 P_{lA}} & \dfrac{\partial^2 \left(A^T A \right)}{\partial Q_{sA} \partial P_{lA}} \\ \dfrac{\partial^2 \left(A^T A \right)}{\partial P_{lA} \partial Q_{sA}} & \dfrac{\partial^2 \left(A^T A \right)}{\partial^2 Q_{sA}} \end{bmatrix}$$

$$\frac{\partial^2 \left(A^T A \right)}{\partial^2 P_{lA}} = 2 * \left(1 + \left(tg(\varphi)_{bef} \right)^2 \right) \qquad (57)$$

$$\frac{\partial^2 \left(A^T A \right)}{\partial^2 Q_{sA}} = 2 * \left(1 + \left(\frac{1}{tg(\varphi)_{stall}} \right)^2 \right) \qquad (58)$$

$$\frac{\partial^2 (A^T A)}{\partial P_{IA} \partial Q_{sA}} = \frac{\partial^2 (A^T A)}{\partial Q_{sA} \partial P_{IA}} = 2 * \left(\frac{1}{tg(\varphi)_{stall}} + tg(\varphi)_{bef} \right)$$

(59)

And the matrix can be written as

$$W = 2 * \begin{bmatrix} 1 + \left(tg(\varphi)_{bef} \right)^2 & \dfrac{1}{tg(\varphi)_{stall}} + tg(\varphi)_{bef} \\[2ex] \dfrac{1}{tg(\varphi)_{stall}} + tg(\varphi)_{bef} & 1 + \left(\dfrac{1}{tg(\varphi)_{stall}} \right)^2 \end{bmatrix}$$

(60)

Determinant for the matrix W is

$$\left(1 + \left(tg(\varphi)_{bef} \right)^2 \right) * \left(1 + \left(\frac{1}{tg(\varphi)_{stall}} \right)^2 \right) - \left(\frac{1}{tg(\varphi)_{stall}} + tg(\varphi)_{bef} \right)^2 =$$

$$= \left(tg(\varphi)_{bef} \right)^2 + \left(\frac{1}{tg(\varphi)_{stall}} \right)^2 + \left(\frac{tg(\varphi)_{bef}}{tg(\varphi)_{stall}} \right)^2 + 1 -$$

$$\left(\left(\frac{1}{tg(\varphi)_{stall}} \right)^2 + 2 * \frac{tg(\varphi)_{bef}}{tg(\varphi)_{stall}} + \left(tg(\varphi)_{bef} \right)^2 \right) =$$

$$= \left(\frac{tg(\varphi)_{bef}}{tg(\varphi)_{stall}} \right)^2 - 2 * \frac{tg(\varphi)_{bef}}{tg(\varphi)_{stall}} + 1 = \left(\frac{tg(\varphi)_{bef}}{tg(\varphi)_{stall}} - 1 \right)^2$$

Finally, the determinant of W is

$$2 * \left(\frac{tg(\varphi)_{bef}}{tg(\varphi)_{stall}} - 1 \right)^2$$

Since the determinant is always positive, the matrix is positive definite for all values of $\tan(\varphi)_{stall}$ and $\tan(\varphi)_{bef}$, except when they have the same value which is not a real case, the critical point is one minima and it is the global optimum.

The values for P_{IA} and Q_{sA} are updated as

$$P_{IA} = P_{IA} + \Delta P_{IA}$$
$$Q_{sA} = Q_{sA} + \Delta Q_{sA}$$

(61)

The initial point used to solve this problem is the values of $P_{lB} = P_A{}^{total}$ and $Q_{sA} = Q_A{}^{total}$.
Fixing a value for β, the remaining values are calculated as

$$Q_{sB} = Q_{sA} / \beta$$

$$P_{lB} = P_{lA}$$

$$P_{sB} = \frac{Q_{sB}}{\tan(\varphi)_{bef}}$$

$$Q_{lB} = P_{lB} * \tan(\varphi)_{bef}$$

$$P_{tB} = P_B{}^{total} - P_{lB} - P_{sB}$$

$$Q_{tB} = Q_B{}^{total} - Q_{lB} - Q_{sB}$$

5.5. Example

The final model of Induction Motors load is finally presented in the
Figure 57 where the superscripts *r s t* account for reaccelerated, tripped and stalled. The induction motor load has been split in fractions related to the reaccelerated, tripped and stalled after the disturbance.

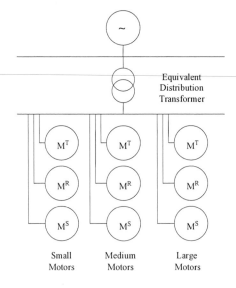

Figure 57. Equivalent model

For the circuit shown in the Figure 58 the identification of motors was performed and the results are shown in Table 7. Each of the 11 KW motors in the Figure 58 represents 5 motors, using the coherent machine option of PSCAD. Inertias for each motor type were set as 0.36sec for the 11 KW motor, 1.09 sec for the 90 KW motors and 10.08 for the 630 KW motor. The 630 KW and 90KW motors represent only one machine. The simulation was performed using random variations in frequency and voltage inputs in the equivalent source shown in the Figure 58 to simulate random variation of the system load. All the motors were started in constant speed mode and after an appropriate time interval were switched to constant torque. Over-current protections were set for each motor, with Ipickup selected as 1.1 times the nominal current and a 0.5 sec time delay.

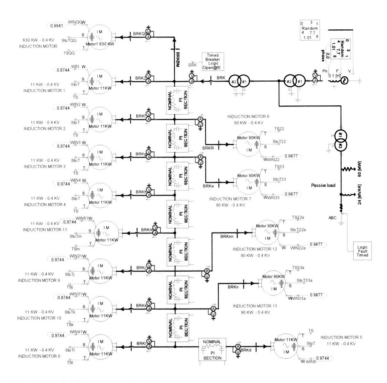

Figure 58.Feeder with induction motor load

Table 7. Motor Identification of the simulation.

Nominal Power	11KW			90KW			630KW		
Inertia	0.18	0.36	0.72	0.54	1.09	2.17	5.04	10.08	20.17
Identified Component %	2%	31%	0%	0%	23%	0%	1%	40%	0%
Identified Component (KW)	29.70	467.03	5.94	0.92	334.13	0.00	11.88	593.82	0.00
Real Component %	0%	33%	0%	0%	24%	0%	0%	42%	0%
Real Component (KW)	0.00	495.00	0.00	0.00	360.00	0.00	0.00	630.00	0.00

The identification of motors was performed using the same procedure than in [38]. The results of the motor identification are shown in Table 7. Results show a good approximation to the real composition of the load, which errors around 10% for the 90KW case.

Figure 59 show the performance of the motors when a fault is applied to the other branch of the grid, inducing a voltage sag around 0.4 pu at the low voltage distribution transformer terminals. Note the value of the sag at Motor 5, which has the longest electrical distance to the source. Figure 59 shows the motors tripping when the voltage increase. The loss of load increased the voltage and the motors tried to reaccelerate, but in some cases the protections tripped and in the cases were the motor is stalled the protections had tripped in a longer time. The voltage at the terminals of Motor 5 has a strong delay to recover, Motor 5 is the last motor to try to reaccelerate and it is stalled. The voltage increase in steps as the rest of the motors are reaccelerated or tripped, but the over-current protection had to wait until the amount of current is strong enough to trip the motor.

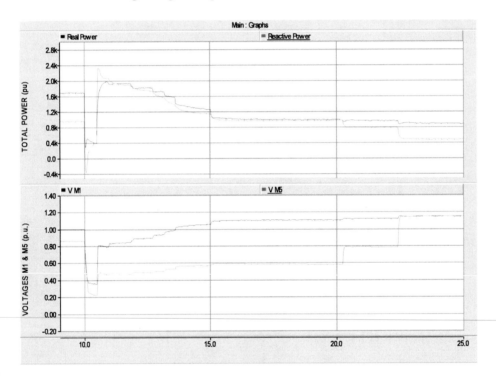

Figure 59. Voltage, Active and Reactive Power

Figure 60 shows the equivalent model for the motors. The selected motors are the same than the identified 11KW, 90 KW and 630 KW with their inertias.

The selection of the fraction of motors to reaccelerate, stall or trip was performed using the information of the severity and tripping index.

From Figure 50 it is clear the 630KW motor would not stall for this voltage sag. Since big motors are commonly used electrically close to the distribution transformer, they don't suffer lower values than the sag at the transformer and for this reason the fraction of tripped 630 KW motor is zero.

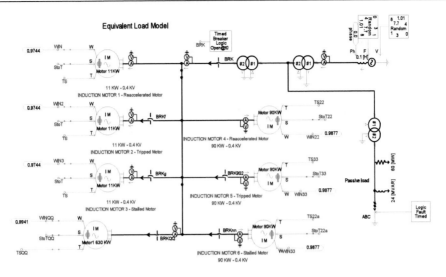

Figure 60. Equivalent Model.

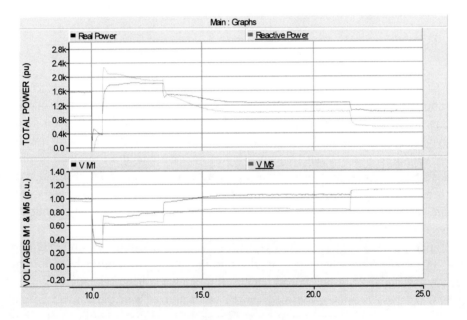

Figure 61. Voltage, active and reactive power from the equivalent model.

From Figure 56 it can be seen that for a voltage dip of 0.4 pu 17 over 28 contactors are tripped representing 60% of the motors. It means that 219kW of 11 kW and 301kW of the 90KW motor type motors will be tripped.

From Figure 50 it can be seen that the 11kW motors would be stalled under the value of the 0.4 pu for more than 0.5 sec. The voltage dip in the simulation is at this limit, which means some of the motors will stall and others not. A supposition of the grid has to be made to obtain the fraction of stalled motors. In this case the information of the voltage at motor 5 is available. The voltage in steady state at the more distant motor is 10% below than the voltage at the transformer. Supposing the voltage is distributed along a radial grid as the square of the voltage, 33% of the motors will be stalled, which is the fraction of stalled motors for this case. From the same figure it can be seen the 90KW motors will not be stalled and the fraction of stalled motors for the 90 KW is zero.

Figure 61 shows the result of the simulation with the fraction of motors selected and using over-current protections to trip the motors. As it can be seen the limited amount of motors means the curves present some steps when the motors tripped. To stall effectively the 11KW motor an impedance was added and the voltage at its terminals is shown.

Comparing Figure 59 and Figure 61 it can be seen that the time of the voltage recovery, the loss of load and the maximums of active and reactive power are correctly represented.

The previous example has shown how data gathered from the small signal model can be used to identify induction motor nominal powers and their inertias. The identified motors were used in the model presented in Figure 60 as pre-fault load.

Severity index and tripping index were used to determine percentages of stalled and tripped motors for the equivalent model. Considerations regarding the voltage dip at the end of the line can be handled easily if the voltage profile of the feeder is known. If the voltage profile of the feeder is not available, the percentages of stalled and tripped motors can be obtained from several faults using the algorithm presented in the previous section.

6. DISCUSSION

The process of load model identification from normal system variations, offers promise of separation of motor load components from total load and the identification of the motor parameters, in particular inertia and nominal power. The quality of identification depends on the degree of frequency variations visible in the system. When there is a low level of frequency variations in particular frequency bands then the quality of the load identification can be affected. When a particular frequency band is rich in electromechanical oscillation modes then the quality of motor load identification is good. This is a fortunate aspect since the main motivation is to quantify the effect of load dynamic parameters on electromechanical modes.

The appropriate assessing of the frequency-voltage relationship which leads to a more accurate small signal analysis. This allows using the peeling method in [30] to extract the impedance load and the constant P load remains through the voltage-P relation.

Following these premises, the load can be characterized and the amount of induction motors can be obtained before the large disturbance. This can not be applied in the post-fault identification, because the quantity of measurements is not enough in 5-30sec to identify the amount of stalled induction motors.

One way around this problem is simulating the induction motor behaviour under major disturbances. Simulations carried out shown strong potential to reproduce major disturbances.

But even when the disturbance can be accurately reproduced, the main objective is to predict an accurate response of the induction motor load component. Otherwise a great number of disturbances need to be known for each feeder to accurately predict the load response.

Since the small signal information is available and if the nominal power is known an acceptable approximation to motor parameters can be easily found and the behaviour of the induction motor can be reproduced. The knowledge of the inertia of the set motor-load is the other variable that modified the behaviour of the motor. Both, nominal power and inertia can be extracted from the small signal model. One of the next tasks is to test the capability of the small signal model to bring accurate values of both parameters using the algorithms already developed at QUT.

The research on Severity Index shows there is a correspondence between parameters of the induction motor and inertia that can be used to predict the stalled or reaccelerated state after a disturbance. The research on Tripping index showed the possibility to predict the amount of contactors that will trip and the thermo-magnetic protection opening in case of stall or high currents due to the reacceleration process performed far away from the source. These indexes shown potential to explain and simulate disturbances and can be developed further if the adequate data and configuration of the grids are available. For industry application the stall fraction for different motor groups, as a function of the dip severity, can be fitted to a number of past disturbances. At any given time the motor load levels at different sizes can be continuously extracted. Knowing the severity of any given fault means that the level of motor loads which stall for each group can be predicted.

7. CONCLUSIONS

This work shows a more accurate way to extract dynamic load models from normal operational data of the power system. Critical aspects pointed out in this work are that the loads need to consider the existence of a feedback and the influence of the voltage magnitude on the relation between frequency and power. An approach to deal with the coloured noise presented at the frequency was presented. From frequency and voltage to power relations the load dynamics can be extracted and motor load portions inferred.

Regarding major disturbances, data gathered in the small signal model can be used to create models for pre-fault load. Severity Index and Tripping index have the potential to predict the comportment of induction motors loads under major disturbances.

ACKNOWLEDGMENTS

The authors acknowledge the support of ARC and Transend. The authors would also thank the QUT High Performance Computing Group for its support.

REFERENCES

[1]   IEEE Task Force on Load Representation for Dynamic Performance, "Bibliography on load models for power flow and dynamic performance simulation," *IEEE Trans. Power Syst.*, vol. 10, no. 1, 523-538, Feb. 1995.

[2]   IEEE Task Force on Load Representation for Dynamic Performance, "Load representation for dynamic performance analysis," *IEEE Trans. Power Syst.*, vol. 8, no. 2, 472-482, May 1993.

[3]   Measurement-Based Load Modeling. Electric Power Research Institute - EPRI Technical report. *December*, 6, 2006.

[4]   Concordia, C; Ihara, S. "Load representation in power system stability studies," *IEEE Trans. Power App. Syst.*, vol. PAS-101, no. 4, 969-977, Apr. 1982.

[5]   Agalgaonkar, P; Kashem, M; Negnevitsky, M. On-line Monitoring and Modelling of Electric Loads for Improving Operational Conditions of Power Systems: *A Literature Review.*, REPORT 18 January, 2008, UTAS, Tasmania, 2009.

[6]   Tanneeru, Sirisha; Mitra, Joydeep; Patil, Yashwant J; Ranade, Satish J., "Effect of Large Induction Motors on the Transient Stability of Power Systems," *Power Symposium,* 2007. NAPS '07. 39th North American , vol., no., 223-228, Sept. 30 2007-Oct. 2 2007.

[7]   Moyano, CF; Ledwich, G. "Load Model: Induction Motors in Large Disturbance" AUPEC 2008–December 2008, Sydney, Australia,

[8]   Williams, BR; Schmus, WR; Dawson, DC. "Transmission voltage recovery delayed by stalled air conditioner compressors," *Power Systems, IEEE Transactions on* , vol.7, no.3, 1173-1181, Aug 1992.

[9]   Rahim, AHMA; Laldin, AR. "Aggregation of induction motor loads for transient stability studies," *IEEE Trans. Energy Convers.,* vol. EC-2, no. 1, 55-61, Mar. 1987.

[10] Undrill, JM; Laskowski, TF. Model Selection and Data Assembly for Power System Simulations, *IEEE Transactions on Power Apparatus and Systems*, Sept. 1982 Vol PAS-101, Issue: 9 pages 3333-3341 ISSN: 0018-9510.

[11] Pereira, L; Kosterev, D; Mackin, P; Davies, D; Undrill, J; Zhu, W. "An interim dynamic induction motor model for stability studies in the WSCC," *IEEE Trans. Power Syst.*, vol. 17, no. 4, 1108-1115, Nov. 2002.

[12] Price, WW; Wirgau, KA; Murdoch, A; Mitsche, JV; Vaahedi, E; El-Kady, MA. "Load modeling for power flow and transient stability computer studies," *IEEE Trans. on Power Syst.*, vol. 3, no. 1, 180- 187, Feb. 1988.

[13] Nozari, F; Kankam, MD; Price, WW. "Aggregation of induction motors for transient stability load modeling," *IEEE Trans. Power Syst.*, vol. PWRS-2, no. 4, 1096-1103, Nov. 1987.

[14] IEEE Task Force on Instrumentation for System Dynamic Performance, "Instrumentation for monitoring power system dynamic performance," *IEEE Trans. Power Syst.*, vol. PWRS-2, no. 1, 145-152, Feb. 1987.

[15] Navarro, IR. "Dynamic power system load and estimation of parameters from operational data," *Ph. D. Thesis,* Lund University, Sweden, 2005.

[16] Pedra, J. "Estimation of typical squirrel-cage induction motor parameters for dynamic performance simulation," *Generation, Transmission and Distribution, IEE Proceedings-* , vol.153, no.2, 137-146, 16 March 2006.

[17] Lem, TYJ; Alden, RTH. "Comparison of experimental and aggregate induction motor responses ," *Power Systems, IEEE Transactions on* , vol.9, no.4, 1895-1900, Nov 1994.

[18]   Thirimger, T; Luomi, J. "Comparison of Reduced-order Dynamic Models of Induction Machines." *IEEE Trans. Power Sys.*, Vol. 16. 119-126, Feb.2001

[19] Pedra, J; Candela, I; Sainz, L. Modelling of squirrel-cage induction motors for electromagnetic transient programs. IET Proc. *Electric Power Applications*, Vol.3, No. 2, 2009, 111-122

[20] Pedra, J. Estimation of typical squirrel-cage induction motor parameters for dynamic

performance simulation *IEE Proc. Gener. Transm. & Distrib.,* Vol.153, No. 2, March 2006, 137-146.

[21] Haque, MH. Determination of NEMA Design Induction Motor Parameters From Manufacturer Data. *IEEE Transactions on Energy Conversion,* Dec. 2008, Vol 23, I 4. 997-1004

[22] Xu Guanghu, Chen Chen and Sun Qu, "The influence of induction motor inertia constant on small-signal stability". *Electric Power Systems Research,* Volume, 74, Issue 2, May, 2005, Pages 197-202.

[23] http://catalog.wegelectric.com. *WEG catalogues of Induction Motors.*

[24] Taylor, LY; Hsu, SM. "Transmission voltage recovery following a fault event in the Metro Atlanta area," Proceedings of the 2000 *IEEE PES Summer Meeting,* July 16-20, 2000, 537-542.

[25] Shaffer, JW. "Air conditioner response to transmission faults," Power Systems, *IEEE Transactions on,* vol. 12, no.2, 614-621, May 1997.

[26] Chinn, GL. "Modeling Stalled Induction Motors," Transmission and Distribution Conference and Exhibition, 2005/2006 *IEEE PES ,* vol., no., 1325-1328

[27] Stewart, V; Camm, EH. "Modeling of stalled motor loads for power system short-term voltage stability analysis," Power Engineering Society General Meeting, 2005. *IEEE ,* vol., no., 1887-1892 Vol. 2, 12-16 June 2005.

[28] Ledwich, G; Parveen, T; O'Shea, P; Moyano, C. "Measurement based Load Model Identification". 84th National EESA *Conference and Exhibition Electricity,* 2008. Brisbane 20-22 August 2008.

[29] Kay, SM. Fundamentals of Statistical Signal Processing: *Detection Theory.,* 1998, Prentice-Hall, Upper Saddle River, NJ.

[30] Bahadornejad, M; Ledwich, G. *"Studies In The OLTC Effects On Voltage Collapse Using Local Load Bus Data"* Proceedings of AUPEC, Christchurch, September, 2003.

[31] Ledwich, Gerard and Zhang, Chuanli (2005) *Analysis of Major Australian Events Using the Angle Measurement System.* In: AUPEC 2005 CD Proceedings: Australian Universities Power Engineering Conference, 25 September - 28 September 2005, Australia, Tasmania, Hobart.

[32] Manitoba HVDC Research Center: 'PSCAD/EMTDC User's *Manual Guide',* Version 4, 2004.

[33] The MathWorks, Inc.: MATLAB 5.3 and Simulink 3.0, Natick, MA, 1999.

[34] http://energyefficiency.jrc.cec.eu.int/eurodeem: EuroDEEM 2000, *European Database of Efficient Electric Motors.*

[35] Lennart Ljung, *System identification: theory for the user,* Prentice-Hall, Inc., Upper Saddle River, NJ, 1986.

[36] Parveen, T; Ledwich, G; Palmer, E. (2006). Model of induction motor changes to power system disturbances. In: *Australasian Universities Power Engineering Conference AUPEC,* 2006, 10th-13th December 2006, Australia, Victoria, Melbourne. (In Press).

[37] Parveen, T; Ledwich, G; Palmer, E. Induction Motor Parameter Identification from Operational Data. In: 2007 Australasian Universities *Power Engineering Conference,* (AUPEC 2007), 9-12 December, 2007, Perth.

[38] Parveen, T; Ledwich, G. Decomposition of aggregated load: Finding Induction motor fraction in real load. In: 2008 *Australasian Universities Power Engineering*

Conference, (AUPEC 2008), December, 2008, Sydney, Australia.

[39] Guasch, L; Corcoles, F; Pedra, J. Effects of Symmetrical and Unsymmetrical Voltage Sags on Induction Machine", IEEE *Transactions on Power Delivery*, Vol. 19, No. 2, April 2004.

[40] Schweitzer, EO. III Zocholl, S.E. The universal overcurrent relay. *Industry Applications Magazine*, IEEE, May/Jun 1996, Vol 2, 3, pages 28-34.

[41] IEEE Power & Energy Society. IEEE guide for the presentation of thermal limit curves for squirrel cage induction machines - 620-1996. *Digital Object Identifier*, 10.1109/IEEESTD.1996.81041. Year: 1996.

[42] Gomez, JC; Morcos, MM; Reineri, CA; Campetelli, GN. Behavior of Induction Motor Protection Due to Voltage Sags and Short Interruptions, Power Engineering Review, *IEEE* , vol.21, no.11, pp.62-62, Nov. 2001.

[43] Morcos, MM; Gomez, M. *J. Determination Of Maximum Time Needed For Motors To Restart After A Perturbation.* C I R E D 19th International Conference on Electricity Distribution Vienna, 21-24 May 2007

[44] El Shennawy, T; El-Gammal, M; Abou-Ghazala, A. Voltage Sag Effects on the Process Continuity of a Refinery with Induction Motors Loads. American *Journal of Applied Sciences*, 6(8), 1626-1632, 2009.

[45] Pohjanheimo, PA. Probabilistic Method For Comprehensive Voltage Sag Management In Power Distribution Systems. Helsinki University of Technology publications in Power Systems, 7, *Espoo* 2003, TKK-SVL-7, October 20th, 2003, Helsinsky, Finland.

[46] Milanovic, JV; Aung, MT; Vegunta, SC. The influence of induction motors on voltage sag propagation - Part I: Accounting for the change in sag characteristics. *IEEE Transactions on Power Delivery*, 23(2), 1063-1071. 2008.

[47] Milanovic, JV; Aung, MT; Vegunta, SC. "The Influence of Induction Motors on Voltage Sag Propagation—Part I: Accounting for the Change in Sag Characteristics," Power Delivery, *IEEE Transactions on*, vol.23, no.2, 1063-1071, April 2008.

[48] Trudnowski, DJ; Donnelly, MK; Hauer, JF. A procedure for oscillatory parameter identification, Power Systems, *IEEE Transactions on*, vol.9, no.4, 2049-2055, Nov 1994.

In: Electric Power Systems in Transition   ISBN: 978-1-61668-985-8
Editors: Olivia E. Robinson, pp. 137-169   © 2010 Nova Science Publishers, Inc.

Chapter 3

ADVANCED WIDE-AREA ANGLE STABILITY AND VOLTAGE CONTROL

A. R. Messina[1]*, R. Castellanos* [2] *, and R. Betancourt*[3]

[1] The Center for Research and Advanced Studies (Cinvestav), Mexico
[2] The Electric Power Research Institute of Mexico, Mexico
[3] The University of Colima, Mexico

ABSTRACT

Wide-area stability and voltage control of large interconnected power systems has attracted considerable interest in the last two decades. Advances in signal processing algorithms, along with continuously growing computational resources and the evolution in protection schemes, are beginning to make feasible the analysis and control of intersystem oscillation problems using measurement-based, wide-area protection systems.

A considerable amount of research has been carried out over the last twenty years to develop advanced, wide-area monitoring, protection and control systems. Conventional stability ad voltage control systems may be inadequate to cope with wide-area stability problems and emergency conditions.

Integrated synchro-phasor measurements are emerging as the backbone for future real-time wide-area monitoring, protection and control systems. The advent of remotely sensed data from satellites greatly extends the scales of time and space at which control can be carried out. Efficient utilization of these technologies, however, poses enormous practical and theoretical challenges that have to be addressed especially under deregulated power system operation.

This chapter discusses the experience in the design and evaluation of advanced wide-area stability and voltage control systems to remove transmission constraints in the system. Examples taken from a large power system show that wide-area Special Protection Systems (SPSs) involving direct tripping of generation (load) and wide-area measurement-based secondary voltage control can maintain system stability and prevent from further system wide spread cascading outages. Control options using both, shunt and series compensation schemes as well as under voltage load shedding are evaluated

Implementation details using WAMS/SCADA systems are also discussed.

WIDE-AREA ANGLE STABILITY CONTROLS

Continuous and discrete wide-area stability controls are being extensively used for advanced angle stability and voltage control. The increasing number of satellite observations and the development of more sophisticated data acquisition systems and signal processing algorithms are forcing utilities to develop real-time monitoring, protection and control systems.

Remotely-sensed data offers the potential to monitor and analyze dynamic patterns and processes that control oscillatory behavior, which are critical to the accuracy of models used to support the development and evaluation of wide-area control schemes.

Based upon planning information and system topology, automatic generator tripping schemes have been applied in coordination with other load and shedding and load restoration programs to maintain transient stability under severe contingencies [1,2]. Highly specialized schemes (SPSs) involve such actions as generation tripping, shunt capacitor switching strategies, load shedding, intertie separation, capacitor insertion or dynamic brake insertion.

Designing remedial schemes is a complex problem in which several factors need to be considered. Central issues that have to be addressed include:

a) Improved near-real time identification/characterization of system oscillatory behavior
b) Coordination of SPSs control actions
c) Determination of the amount and location of load (generation) to be shed and the optimum time delay to initiate the load (generation) shedding
d) Development of control logic and the mechanism of activation of the schemes
e) Demonstration of real-time system performance of the proposed methods

These issues are discussed separately below.

Adaptive Wide-Area Special Protection and Control Systems

Special stability controls are broadly defined as discontinuous or continuous control actions for the stabilization of electromechanical and voltage dynamics [3]. Highly specialized special protection systems may be local, system wide or involve interconnected power systems as discussed in [4]. Control structures, in turn, may be centralized, decentralized, local or hierarchical.

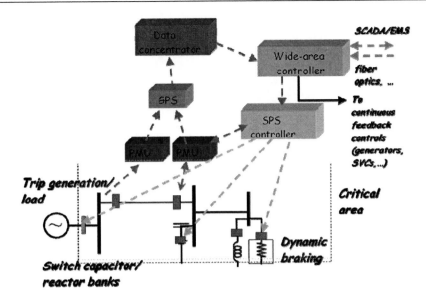

Figure 1. Structure of power system synchrophasor-based wide-area closed-loop control

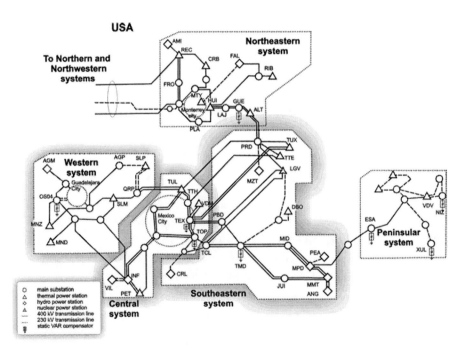

Figure 2. Simplified one-line diagram of the Mexican 400/230 kV transmission system showing major regional systems and transmission and generation resources

Figure 1 shows an overview of a wide-area control system involving continuous and discontinuous control actions. Both, hierarchical and multi-layer configurations have been proposed in the literature [5].

Comprehensive system protection against extreme contingencies requires the solution of three major problems: a) monitoring of key system variables, (b) Identification of the mode of instability/islanding, and, (c) the development of countermeasures to aid system stabilization

[6]. The modeling and control of such post-disturbance phenomena raises a number of complex control issues that have to be addressed in real or near-real time.

Adaptive wide-Area Control (Wide-Area Response-Based Control)

Until recently, special protection systems were usually designed to protect the system against loss of stability following predefined system contingencies. Adaptive control systems based on the time evolution of key measured variables are needed to make full utilization of the transmission network especially under severe variations in system behavior as well as to improve selectivity and accuracy of the remedial control actions.

Most of the recent efforts at controlling wide-area phenomena have involved advanced feedback control schemes using wide-area measurements designed to match the evolving power system characteristics [4]. Adaptive wide-area controls, however, need to be globally coordinated to avoid conflicting interactions and require dedicated communication and computer systems [7].

The following sections describe the experience in the planning and evaluation of wide-area stability controls in the Mexican interconnected system. Both, continuous and discrete types of special stability controls are considered. The focus is concentrated on the analysis and control of transient post-fault oscillations which result from critical system contingencies.

IMPLEMENTATION OF WIDE-AREA STABILITY CONTROL SCHEMES USING SYNCHRO-PHASOR DATA

The Mexican Interconnected System

The Mexican interconnected power system is a large system with complex dynamics [8,9] that encompasses the interconnected operation of seven regional systems. The seven areas are North (N), Northwestern, Northeastern, Central, Western, Southeastern (SE) and Peninsular (P). The peak load of this system is about 30000 MW.

Major interconnections include the 400-kV north-south interface and the 230-kV southern-peninsular interface which are limited by small-signal stability. Dynamic instability and islanding for loss of generation or loss of import disturbances has become a concern in the MIS [8].

A highly simplified one-line diagram of this system is shown in Figure 2 depicting major regions of the MIS and transmission paths. Because of its sparse interconnection, large system contingencies result in severe generation-load imbalances often leading to undamped oscillations and uncontrolled system separation. The recent interconnection of the MIS with the Central America system as well as the increase in the number of independent producers and the introduction of wind energy and other renewable generation has resulted in more variable and uncertain operating conditions often imposing additional strain on critical interfaces.

Description of Base Cases

Numerical simulations of the dynamics of the test system are carried out to investigate the efficacy of a wide-area control system based on synchro-phasor data to enhance system-wide performance. Simulations are made on two well-established and calibrated dynamic models:

a) A full 7-area power system model
b) A 130-generator, 6-area dynamic equivalent system model in which the major inter-area modes and system characteristics are preserved

The detailed seven-area system model contains 377 detailed generators, 12 SVCs and 230 undervoltage load shed relays, representing the bulk 400/230/115 kV Mexican network is derived to assess the fundamental nature of inter-area oscillations and the evaluation of remedial measures.

Details of the system representation and component modeling as well as model additions and data modifications are presented in the next sections.

Oscillatory Stability Monitoring

Due to its sparse nature and special characteristics, the MIS is prone to wide-area instability phenomena both under stressed operating conditions as well as under various contingency scenarios. The system has several oscillatory modes that require careful monitoring [9]. Some previous transient small-signal stability studies have shown that the most extreme northeast portion of the transmission system is sensitive to transmission system stress in the north-south interface.

A large-scale software package was used to identify individual critically stable modes, as well as circumstances resulting in low damping. Modal analysis is used to design the best load (generation) shedding schemes to increase system stability margins following major contingencies and provides critical information needed in the identification of vulnerable system regions.

Table 1 lists the main characteristics of the three slowest modes in the test system. Of specific concern within the MIS, the 0.32 Hz interarea mode 1 represents an undamped oscillation involving machines in the south systems (Central, Western Southeastern ad Peninsular areas) swinging in opposition to the group of machines in the north systems (Northern and Northeastern areas).

Detailed system simulations have been conducted to identify the key system parameters for monitoring, and initiating the SPSs. Simulations are made for representative contingency conditions, as well as for reference base case conditions.

The worst-case dynamic response of the system occurs when circuits close to the critical interfaces are lost. Based on previous studies dynamic response voltage profiles at critical system buses were recorded for major disturbances. Also of concern are post-contingency oscillations caused by the loss of major generation resources.

Table 1. Low-frequency inter-area modes.

Mode	Eigenvalue	Frequency (Hz)	Damping (%)	Swing Pattern
1	-0.1143±j2.040	0.325	5.48	North systems vs. Peninsular and Southeastern systems
2	-0.1376±j3.269	0.520	4.20	Peninsular vs. Western system
3	-0.0861±3.948	0.628	2.18	Northern vs. Northeastern system

Figure 3 is an example of the time response of a critical bus voltage to the trip of major transmission and generation resources.

As can be seen in this plot, major system disturbances result in undamped system operation and islanding. Extensive simulation show that the time frame in which dynamic instability can occur is up to several seconds. This is an important factor since signal processing algorithms can be efficiently applied to extract modal properties from relatively slow instability phenomena.

The Mis Wams System

The MIS has a comprehensive Wide-Area Measurement System (WAMS) that provides critical information on system behavior [10]. This system has provided more thorough monitoring of the system and complements the existing Supervisory Control and Data Acquisition (SCADA) system.

At present, data are sent to ten Phasor Data Concentrators (PDCs) for analysis, protection and control purposes [11,12]. Each PDC unit has the potential for providing real-time observations of power system behavior across a broad region of the power system. Measurement sites are well distributed and they can be used to characterize global phenomena across the Mexican system. Reference [12] provides details on the location of main regional centers and the distribution of data concentrators.

In addition to the WAMS, advanced warning systems based on Hilbert and Prony analysis are being developed with the ability to provide real-time monitoring of critical system oscillations. These systems have three components: sensing, processing and decision.

The following sections examine the experience with special protection schemes. Two wide-area special protection schemes are being evaluated and tested. The first is based on a response-based automatic generation dropping scheme. The second makes use of under-frequency load shedding schemes.

They also discuss exploratory studies aimed at using the existing WAMS/SCADA system to enhance power system stabilization. First, a brief review of the existing SPS is presented.

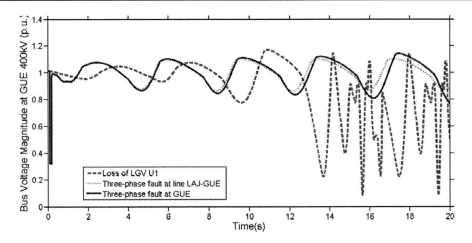

Figure 3. 400-kV Guemez (GUE) response to critical contingencies

Operating Experiences with SPSs and Functional Needs

Special protection systems (SPS) are increasingly being utilized in several areas of the Mexican interconnected system (MIS) to enhance voltage control and system dynamic performance. These include extensive use of direct load shedding schemes, generator tripping schemes, controlled disconnection of lines, and single phase reclosing schemes, among other SPSs [8]. The implementation of this logic is fully automated and utilizes the existing SCADA system.

Of particular interest and importance, generator tripping based on local detection of severe disturbances has been used to enhance transient stability. The activation of controls is dependent upon generation level and system configuration and the status of major static VAR compensators. This class of protection can be extremely complicated due to the large number of outage and disturbance conditions which need to be detected and needs to be coordinated with other frequency (voltage) schemes and static VAR compensation support. This control strategy has proved to be very useful to maintain first-swing stability.

In recent years, the onset of system instability has become more complex, often involving interactions between major system modes. Post-fault transient oscillations have become more common following the loss of major system elements and may result in uncontrolled system separation.

The sections that follow compare the experience in the planning of response-based generation and load tripping schemes for the MIS. Challenges in the design of these systems are also discussed.

Response-based Generator Dropping Scheme (GDS)

Several generator tripping schemes have are in use in the MIS to increase power transfers and enhance first-swing stability [8]. Conventional generator dropping schemes, however, have several disadvantages and are not adaptive to the changing system conditions. As the system grows in complexity, a need is felt to address new and future dynamic phenomena.

In this research, a centralized wide-area GDS scheme is being tested and evaluated to stabilize critical inter-area oscillations in the north systems, which complements other control strategies. Successful on-line implementation of the proposed on-line GDS requires the solution of several interrelated problems:

- The identification of deteriorating system conditions from PMU measurements
- The identification of the critical generators,
- Selectivity to identify the optimum number of generation to be shed
- The timing of the control action and the associated initiation logic

By combining off-line studies with on-line information, a comprehensive wide-area generator dropping control is being evaluated and tested that reduces overtripping.

Requirements for an on-line generator dropping control

As discussed above, efficient design of generator dropping controls requires the solution of two main problems: sensitivity and selectivity [14]. Sensitivity, as discussed in this paper implies that adequate generation amount be dropped to maintain the post-disturbance system stability. Selectivity implies that unnecessary generation drop should be avoided.

In general, local measurements are known to lack observability of wide-area angle transients. As a consequence, it is difficult to achieve both, high sensitivity and selectivity for a generator dropping control which is based on local measurements only. Synchronized wide-area measurements may improve network observability, and provide a better visualization of the system time-varying dynamics.

Inter-area oscillation monitoring

An automated wide-area generation tripping scheme requires both real-time detection of angle instability and the identification of the mode of instability or breakdown. For typical contingencies resulting in the inter-area mode phenomenon, it is required to estimate the damping and frequency of the observed oscillations in real (near-real) time.

As suggested above, the choice of measurement location is important to improve wide-area visibility. In the proposed on-line generator dropping scheme, system dynamic behavior is monitored at key system locations and the instantaneous damping and frequency of the dominant oscillations are extracted using Hilbert (Prony) analysis [15].

Accurate short-term prediction of the modal properties is central to the proposed intelligent wide-area measurement and control systems. Once the modal characteristics are obtained, off-line information is used to identify the critical generators having the largest participation in the observed oscillation.

Selectivity of generator dropping controls

Selectivity is a first step to be addressed in the implementation of an automated GDS scheme. As a solution to this problem, the use of GPS-synchronized measurements has been recently suggested for improving the selectivity of generator dropping controls [14].

Towards this end, a centralized *decision algorithm* has been developed for on-line analysis of inter-area oscillations. The algorithm performs four basic functions:

1. Detects the impending oscillations
2. Extracts and tracks instantaneous frequency and damping
3. Computes a control strategy, and
4. Issues command actions to selected controllable devices

Such an approach is useful for the analysis and control of system-wide phenomena in which the most effective control action is not limited to a single machine or generating unit.

Figure 4 shows a conceptual view of the proposed generator dropping scheme. With appropriate modifications, these procedures can be used to develop a response-based automated load shedding scheme as discussed later in this document.

The decision algorithm monitors key system variables and compares the instantaneous damping with a pre-determined low-damping threshold. If the damping drops below the threshold for a certain time period, the algorithm issues appropriate generator (load) tripping commands to selected generating plants in the system. Action is terminated when the damping of critical system variables falls below a damping threshold value. With this approach, the control action can adapt to system varying conditions and can handle a variety of disturbances in the system in a self-observing, self-correcting manner.

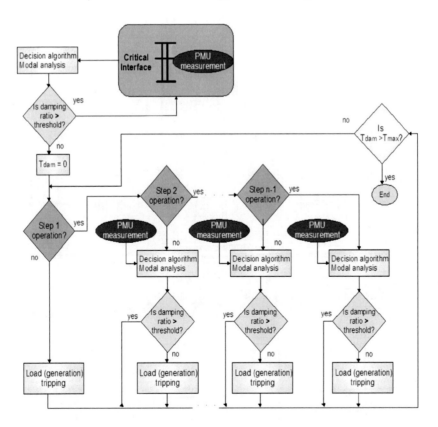

Figure 4. Proposed generator (load) tripping scheme.

The effect of control action, the amount of generation to be rejected and the amount of dropped load on system behavior is investigated. An efficient numerical implementation is also developed based on advanced signal processing algorithms. Based on early experience with the use of advanced signal processing algorithms, fast extraction of mode damping and frequency is accomplished through Hilbert/Prony analysis of measured data. These methods can accurately extract time-varying frequency and damping by processing only a small number of measurement samples. A detailed description of the former algorithm can be found in [15].

    We next discuss some implementation details and discuss practical issues in the design of wide-rea angle stability controls.

Experience with the design and testing of generator dropping schemes

    A wide-area generator shedding scheme is being evaluated and tested to stabilize inter-area oscillations in the northern systems of the MIS. Operational experience and simulation results show that key generator outages in the south systems result in an inertial response in the north systems in which energy is transferred to the south system through the north-south intertie. Extensive results show that critical contingencies excite the slowest inter-area mode at 0.32 Hz. The post-fault behavior is poorly damped especially under stressed operating conditions and becomes unstable after a few seconds. Such inter-area mode instability limits north-south power transfers and eventually results in uncontrolled islanding of the MIS system.

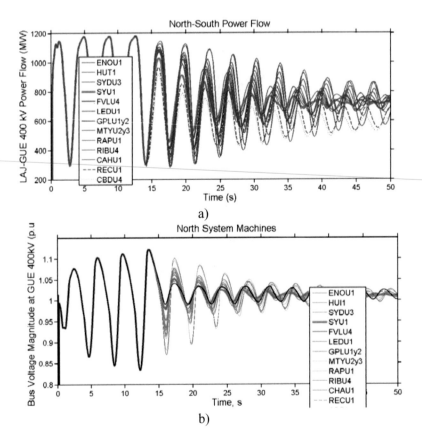

a)

b)

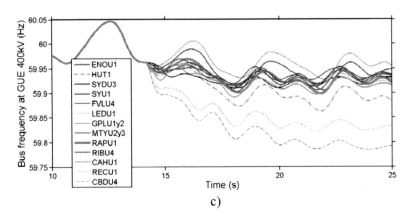

c)

Figure 5. Effect of generator dropping on the damping of critical inter-area oscillations

Table 2. Selected machines for wide-area generator dropping in the Northern system.

Machine(s)	Area	Generated Power (MW)
CBD-U4	Northeastern	350
REC-U1	Northeastern	300
HUT-U1	Northeastern	225
CAH-U1	Northeastern	165
LED-U1	Northern	160
SYC-U1	Northern	150
RAP-U1	Northeastern	150
RIB-U4	Northeastern	144
MTYU2y3	Northeastern	130
ENO-U1	Northern	130
FVL-U4	Northern	125
SYD-U3	Northern	120
GPP-U1y2	Northern	102

Figure 5 illustrates the effect of generator tripping on the damping of the critical inter-area oscillation. For clarity of illustration, Table 2 gives a list of all generators eligible for tripping showing their nominal capacity and location. These generators are chosen based on their influence on system behavior and on their level of integration in the existing SCADA system.

Implementation of this scheme can prevent frequency instability and islanding and result in increased power transfers through critical interfaces. Simulation results in Figure 5 indicate that synchronous machines with high participation factors in the inter-area mode are good candidates to be tripped in order to stabilize the oscillation.

For the particular contingency scenario studied, machine SYU-U1 located in the northern system appears to be the most effective alternative to system stabilization. This has been also observed for other extreme contingencies occurring in the system. It should be stressed that this machine is located nearly 1000 km away from the geographical boundary of the 0.32 Hz north-south mode.

Selectivity of generation dropping schemes

A second issue to be addressed is that of selectivity. In the existing GDS in the MIS, generation dropping is fixed in advance according to load and topology conditions and fault detection.

Selectivity can be improved by recognizing the severity of the fault and monitoring the post-fault dynamics to determine the instantaneous amount of generation dropping required for a disturbance.

Based on the damping of the post-disturbance oscillation, the centralized controller issues a trip signal to the highest priority generator connected at the occurrence of the fault. Figure 6 shows some selected simulations illustrating the impact of generator tripping at CBD and SYD units on system damping. In most cases, direct tripping proper amount of generation can maintain system stability and eliminate the need for generation schedule curtailment or scheduled sequential tripping of multiple generators. The studies demonstrate that the larger the amount of generation real power to be shed, the larger the damping ratio of the post-disturbance oscillation.

Using the results of the previous analysis, sensitivity studies were conducted to calculate the amount of required generation tripping.

Key observations are:

- One-shot generator tripping is an effective measure to enhance damping
- For slow phenomena, damping of post-disturbance oscillations can be improved through generator rescheduling or partial (individual) tripping of units in a plant
- The analysis suggests that an optimum solution may determined by defining a threshold value

SPS steps

A second important factor in the design of response-based wide-area load shedding schemes is the selection of generation trip logic.

Once the first generating unit is disconnected, the damping and frequency of the oscillation has to be monitored. If the damping of the oscillation drops below a critical level, a trip signal is sent to the selected generators to initiate the control action. The monitoring system monitors specific system conditions every few seconds and determines the damping of the observed oscillation; a central controller using both off-line and online information identifies the amount, type and location of generation to be disconnected and checks that no load-generation unbalances and overvoltages are generated. The amount of additional generation tripping needed to stabilize the system is then determined. By looking at the damping of critical signals as a function of time, the algorithm decides if the control action continues and the amount and distribution of generation to be shed.

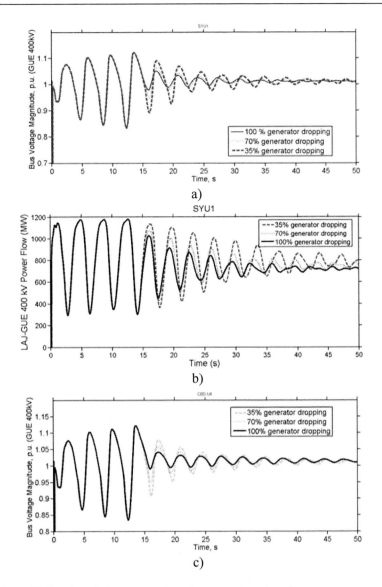

Figure 6. Effect of full and partial generator dropping on system damping

The above approach was used to determine the minimum amount of generation tripping required to maximize the bulk power system voltage security in a steady state sense under post-contingency conditions.

Figure 6 illustrate these ideas. The analysis demonstrates that accurate estimation of system damping conditions may result in effective system stabilization of transient events. The simulation plots also show that the damping criteria are met.

Time delay to initiate generator tripping

A second important factor in the implementation of the proposed method is the delay time required to initiate the tripping action for the first generator (load) tripping step. In addition, it is also important to spread generator tripping over a large number of steps (generation blocks or generating units). This has to be performed on-line to minimize overtripping.

Simulation results below show that the larger the amounts of generator power to be tripped, the larger the delay time needed to trip o disconnect generation. This may be an important factor because the SPS can initiate tripping after a larger time delay to compensate for time delays in communication channels.

Other factors to be considered include:

- Measurements of voltage and current phasors
- Communication delays
- Proper time needed to assess the mode of instability/breakdown
- Actuation delays and transfer trip telecommunications for generator (load) tripping
- Control delays

Table 3. Nominal generated power of the North region machines

Activity	Comments
Time delays effects of measurement equipments, breakers and other effects	Effects not fully accounted for in the present simulation studies
Real-time information processing/ control logic for generator dropping	Two-cycle intervals needed to produce reasonable damping estimates
Trip signal transmission to load breakers/Communication delays	Dependent on the communication channels*
Control time delays	Typical values used in simulations

* Fiber optics and microwave communication channels in the MIS

Time delays associated with real-time information processing, and communication and actuation delays must be considered as indicated in Table 3 and discussed in Ref. [16].

These delays are highly dependent on the nature of communication links and other factors and are, to some extent, uncertain and unknown.

Wide-Area Under-Frequency Load Shedding Scheme

The MIS has an underfrequency load shedding scheme that is employed to minimize frequency variations following major contingencies [8]. Suitable load shedding sites for the SPSs are selected based on extensive studies considering various expected operating scenarios. Typical frequency thresholds for load shedding are: 59.7 Hz, 59.2 Hz, and 59.6 Hz.

Extensions and refinements to the existing load shedding scheme are evaluated here to analyze the capacity of load rejection to control post-disturbance oscillations.

Similar to the case of GDSs, efficient online implementation of wide-area under-frequency load shedding schemes requires the identification of the most suitable loads to be shed, the required amount of load shedding and the number of steps for on-line load disconnection.

By looking at damping as a function of time the first load block is switched off. If damping does not reach the desired post-disturbance level, the next load block is switched off. The decision algorithm issues switching actions to the most effective load locations until a secure dynamic condition is obtained.

In the present phase of this research, the location and size of the load blocks to be shed is determined in two steps. In the first step, modal analyzes similar to those described later in this chapter are conducted to determine the sensitivity of critical modes to load variation. In the second step, extensive post-disturbance large-scale transient stability simulations are undertaken to determine the timing of the control action as well as the amount of the load shedding blocks require to stabilize the system for a given contingency.

The solution is not unique since each contingency may result in a different dynamic pattern. Alternatives to the above approach based on the use of more sophisticated techniques are being actively investigated.

This is an open area of research and deserves active investigation.

Sensitivity analyzes

Sensitivity studies conducted on critical contingency conditions indicate that the effectiveness of load shedding schemes to stabilize critical oscillations is heavily dependent on the amount and location of load to be shed. The first issue is to determine how large the load to be shed can be. Having determined the amount of load the next issue is the timing of the control action. Preliminary results from this comparison indicate that a tradeoff should be made between the size of the load and its capability to effectively stabilize the system.

To illustrate, Figure 7 shows plots of the power flow at the north-south intertie in the MIS following the loss of various generation/transmission resources. Comparison of Figures 7a) and 7b) shows that automated load shedding of suitable load blocks can increase transmission capacity on the north-south intertie. This also demonstrates the need for re-evaluating system dynamic characteristics using time-varying signal processing algorithms.

A ranking of the effectiveness of the proposed load shedding schemes is presented below in Table 4. For reference, the geographical location of the load is listed along with the tripping required to stabilize the system.

Table 4. Ranking of loads to be shed.

Substation	Location (Area)	Amount (MW, MVAR)	Tripping Time (s)
PNT	Central	74, 17	14
TAX	Central	84, 23	
CHA	Central	97, 30	
NET	Central	77, 23	15
KCR	Central	79, 27	
VIC	Central	95, 34	
XOC	Central	96, 29	
REM	Central	97, 26	
MAG	Central	104, 34	
TUL	Central	88, 20	18
IZT	Central	122, 39	
MAD	Central	129, 40	
ECA	Central	104, 41	19
GUE	Northeastern	121, 40	
GUE+TUL	Northeastern and Central	209, 60	23
PNT, IZT.MAG	Central	300, 90	

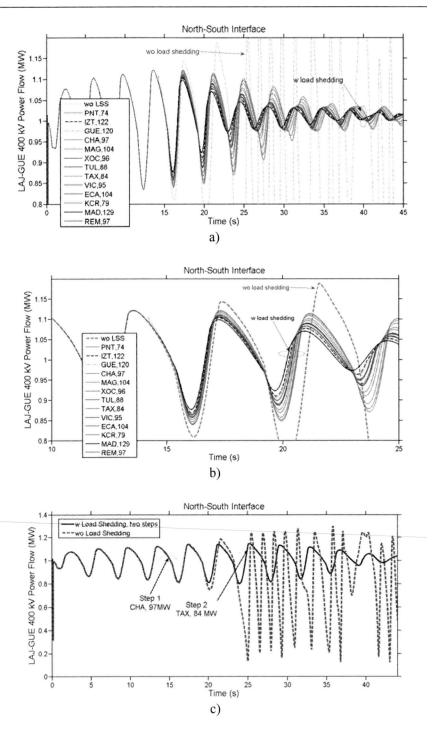

Figure 7. Effect of direct and sequential load shedding on system damping

The results summarized in Table 4 give a clear indication of the relative performance of the load shedding schemes and show that the location of load has an important effect on the ability of LSS to enhance damping of transient oscillations.

The following points may be noted here:

(a) The best response from shedding load is found far from the critical interface (PNT, TAX and CHA), in this case in the central system which has the largest load concentration in the system.

(b) The timing of the control action is a second aspect to be considered. It is interesting to note that loads closer to the interface (GUE,TUL) enable larger trip times but usually require large amounts of load shed to stabilize the system.In general, the larger the tripping time, the larger the amount of load to be shed to achieve a similar stabilization objective

(c) Simulation results in Figure 8, on the other hand, show that the best locations to shed load are associated with robust buses in the voltage stability sense, i.e. those associated with less voltage variations when the load is shed. These observations are in good agreement with sensitivity results for typical operating conditions studied.

Simulation results in Figure 8 demonstrate that timely activation of the wide-area stability control is critical to the efficacy of the proposed scheme to control post-disturbance phenomena. For the specific contingency scenario studied, the shedding action is taken about 13 seconds after the event occurs.

Another important observation results from an analysis of the impact of switching blocks of load on system behavior. From Figure 9, it can be observed automated sequential shedding of blocks of load can effectively stabilize the system.

WIDE-AREA VOLTAGE CONTROL

Wide-area area voltage control of large interconnected systems has attracted considerable interest in the last two decades [17,18]. With significant progress in control techniques, signal processing and measuring, the successful implementation of wide-area voltage control technology is becoming a reality.

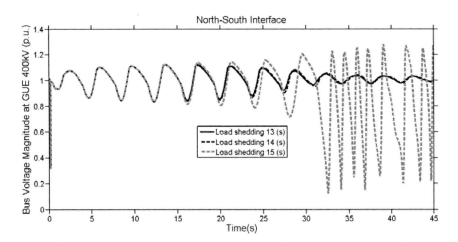

Figure 8. Impact of time delays on the ability of generator dropping controls to stabilize critical oscillations

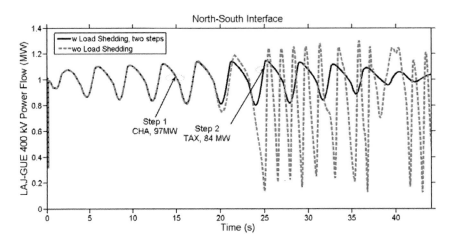

Figure 9. Number of load shedding blocks in the LSS scheme

Flexible ac transmission devices placed at key system locations can contribute to voltage stability and angle control of stressed power systems. Static VAR compensators in particular have the ability to support voltage control and enhance system transient stability operating flexibility [19-21].

In the recent past there has been renewed interest into the application of voltage-VAR controls to maximize reactive power reserves. Voltage-VAR control can contribute to overall power system angle and voltage stability and result in improved system operation and security [22,23]. Coordinated voltage control may also play a role in preventing voltage collapse and benefit angle stability. The increasing size and complexity of power systems, however, makes voltage and angle control problematic [24-26].

The modeling and control of advanced voltage-VAR control strategies raises a number of complex issues. Experience with the application of secondary voltage control in power systems in European countries shows that coordination of reactive sources may result in enhanced system control and reliability. To avoid undesirable interactions, voltage control at the various levels should temporally and spatially independent. This requires splitting the system into non-interacting zones in which voltage control is controlled individually.

With the advent of advanced and more accurate measurement techniques and new and faster control devices, the problem of voltage and reactive power control is becoming an increasingly critical concern. Large scale coordination of reactive power sources is very challenging due to the large number of control characteristics, the location and type of controllers, and the characteristics of each device. Issues such as reserve capacity control and the efficient utilization and coordination of reactive power sources have to be addressed to achieve fast voltage control and keep the capacitive output margin against system contingencies.

This section discuses the experience in the modeling and simulation of secondary voltage control schemes for the MIS. Emphasis is being placed on the ability of wide-area control of several Static VAR Compensators (SVCs) to improve both power system angle and voltage stability.

A systematic methodology for the analysis, synthesis and coordination of reactive sources is proposed. The proposed procedure consists of three main steps: (a) the identification of modal structure using the notion of power oscillation flow, (b) the identification of generators and SVCs participating in the critical modes, and (c) the coordination of reactive compensation devices. The method is suitable for large-scale applications and can be used to coordinate multiple available reactive compensation devices including SVCs, synchronous condensers, and generator excitation systems.

The design methodology is demonstrated on a six-area model of the Mexican interconnected system in which several SVCs are used to control system voltage. Results show that properly coordinated reactive power sources may have an important impact on system dynamic behavior.

Operating Experience with SVCs

Static VAR compensation (SVCs) support has been extensively used in the MIS to enhance dynamic performance as well as to provide continuous voltage control on the bulk 400/230 kV transmission system [27]. Benefits from secondary voltage control include improved voltage stability, improved system reliability, and enhanced small signal and transient behavior.

Because of their strategic location near critical nodes, SVCs can also influence overall system damping. Table 5 gives summary information on some major SVCs selected for analysis. Large SVCs include the SVCs at the 400 kV substations TMD, TEX, TOP, SVCs the 230 kV substation ESA, and at the 115 kV substation XUL.

A simplified one-line diagram of a typical SVC used in various recent applications is shown in Figure 10 identifying key equipment components. Other SVCs in the system have similar configurations [28].

These SVCs are normally used to provide maximum reactive power reserve to prevent loss of stability following major contingencies.

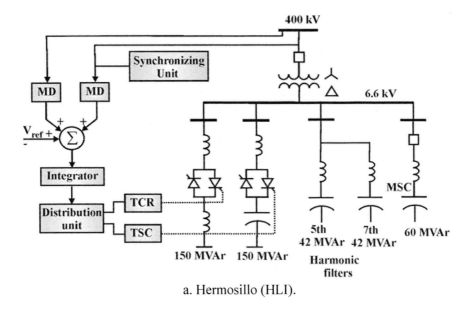

a. Hermosillo (HLI).

Figure 10. Simplified one-line diagram of the SVC at Temascal

Table 5. Summary of the SVCs

SVC Location	Rating (MVAr)	System Voltage	Area	Main Purpose
ESA	-50/150	230	Peninsular	DVC/PSD
XUL	-20/40	115	Peninsular	DVC
TEX	-90/300	400	Central	DVC
TOP	-90/300	400	Northeastern	DVC/PSD
TMD	-300/300	400	Southeastern	DVC

* DVC – Dynamic voltage control, PSD-Power system damping

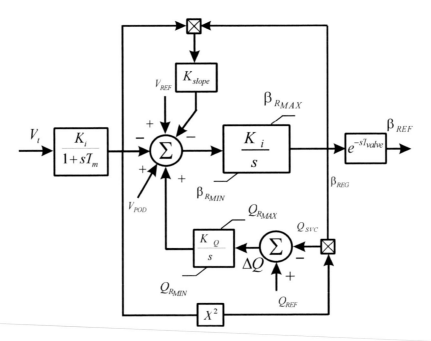

Figure 11. Block diagram of the V-Q control.

Voltage-VAR Controls

Modern Secondary voltage-VAR controls with external capacitor switching have been recently implemented in various SVCs to allow the SVC to control voltage while maintaining the reactive power reserve against contingencies.

Figure 11 shows a schematic of the voltage and reactive power control loop used in some existing SVCs. The primary control loop includes a measuring device (MD), and an integral controller. The adjustable slope is obtained by feedback of the SVC reactive current. In addition, some SVCs are equipped with a slow Q-controller (AQR) designed to maintain the output of the SVC at a predetermined reactive power setpoint (Q_{ref}). Through the modification of the input reference values of the AVRs, V_{ref}, , the reactive reserves can be efficiently optimized on a real-time basis.

The AQR control function compares the reactive power set point Q_{ref}, to the reactive power output, Q_{svc}, and derives a difference value ΔQ. The difference value is then processed through an integral controller to obtain the voltage command for the AQR output. Coordinated capacitor bank control is used in some devices to assure maximum dynamic reserve of the SVC for system disturbances.

In current practice in the Mexican system, the AQRs are tuned individually which prevents coordinated wide-area voltage control.

Wide-Area Reactive Power Control - Secondary Voltage Control

Secondary voltage control has recently emerged as a viable tool for enhancing system operation and reliability. Existing secondary voltage control schemes rely on the decomposition of a large power system into theoretically non-interacting areas, with the voltages within each area being independently and hierarchically controlled.

Figure 12 illustrates the structure of existing hierarchical voltage control schemes [20]. Coordination of regional voltage and reactive power controls is achieved by slow control of the AVR system of power plants, SVCs, and synchronous condensers. Slow secondary voltage control (AQR) is usually an order of magnitude slower that the primary voltage control system and is normally used to compensate for slower voltage variations.

In present implementations, the AQR adjusts the reactive power of selected SVCs and machines to control the voltage at specific points in the network called pilot buses. Key to this approach is the identification of geographical regions or areas showing a common behavior. As discussed by several authors, the selection of pilot buses is a complex nonlinear problem. In practice, system decomposition is performed based on steady-state analysis of the powerflow solutions.

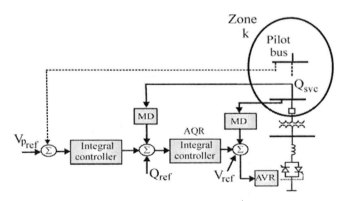

Figure 12. Hierarchical control structure for the SVCs.

This, however, precludes investigation of the role of wide-area voltage control on system dynamic performance. References [24,26] discuss current design methodologies.

When wide-area angular stability is also of concern, especially in sparsely interconnected power systems, the small signal angle behavior has to be taken into account. This is the focus of this research.

In what follows, a modal-based approach to determine reactive power areas that incorporates both voltage and angle stability information is proposed. The method is successfully tested on the Mexican interconnected system.

Analysis Method

A modified large-scale small signal stability program is used to determine voltage control areas, the associated pilot buses for reactive power support, and existing reactive power sources having a strong influence on system behavior.

Following Messina et al. and Kundur [29,30] let the small-signal system behavior be expressed in the form

$$\dot{\mathbf{x}} = \mathbf{A}_d \mathbf{x} + \mathbf{C}\Delta\mathbf{v} + B_d \Delta\mathbf{u}$$
$$\Delta \mathbf{i}_d = \mathbf{W}_d \mathbf{x} + \mathbf{Y}_d \Delta\mathbf{v}_d$$
$$\Delta \mathbf{i} = \mathbf{Y}_{bus} \Delta\mathbf{v} \tag{1}$$

in which

$$\mathbf{y} = \begin{bmatrix} \mathbf{C}_d & \mathbf{C}_n \end{bmatrix} \begin{bmatrix} \Delta\mathbf{x} \\ \Delta\mathbf{v} \end{bmatrix} \tag{2}$$

Where
- x   states associated with each device
- v_d voltages ate the device terminal with the network
- I   vector of current injections into the network buses
- V   Vector of network voltages

Defining $\Delta \mathbf{i} = [\Delta \mathbf{i}_d \quad \mathbf{0}]^T$, $\Delta\mathbf{v} = [\Delta\mathbf{v}_d \quad \Delta\mathbf{v}_d]^T$, $\Delta\mathbf{v} = [\Delta\mathbf{v}_d \quad \Delta\mathbf{v}_d]^T$, and

$$\mathbf{W} = \begin{bmatrix} \mathbf{W}_d \\ \mathbf{0} \end{bmatrix}; \quad \mathbf{C} = \begin{bmatrix} \mathbf{C}_d & \mathbf{0} \end{bmatrix}; \quad \mathbf{Y} = \begin{bmatrix} \mathbf{Y}_d & \mathbf{0} \end{bmatrix}$$

the composite system model would then have the form

$$\begin{bmatrix} \dot{\mathbf{x}} \\ \mathbf{0} \end{bmatrix} = \begin{bmatrix} \mathbf{A}_d & \mathbf{C} \\ \mathbf{W} & \mathbf{Y}_d - \mathbf{Y}_{bus} \end{bmatrix} \begin{bmatrix} \mathbf{x} \\ \Delta\mathbf{v} \end{bmatrix} + \begin{bmatrix} \mathbf{B}_d \\ \mathbf{0} \end{bmatrix} \mathbf{u} \tag{3}$$

Equation (3) is used to determine critical system modes and the associated modal information. Extensions to this basis formulation to identify voltage control areas strongly associated with a swing mode are discussed next. For the operating conditions considered, the open loop system has a dimension of 1365.

Critical Bus identification using modal voltages

Useful insight into the effect of voltage control on small signal stability can be found from the analysis of modal voltage solutions.

Solving for the bus voltage deviations in (1) in terms of the state solution yields

$$
\begin{bmatrix} \mathbf{Y}_{bus\,dd} - \mathbf{Y}_d & \mathbf{Y}_{bus\,dd} \\ \mathbf{Y}_{bus\,Ld} & \mathbf{Y}_{bus\,LL} \end{bmatrix} \begin{bmatrix} \Delta \mathbf{v}_d \\ \Delta \mathbf{v}_L \end{bmatrix} = \begin{bmatrix} \mathbf{W}_d \\ \mathbf{0} \end{bmatrix} \mathbf{x} = \mathbf{W}\mathbf{x}
\tag{4}
$$

and

$$
\begin{bmatrix} \Delta i_d \\ \mathbf{0} \end{bmatrix} = \begin{bmatrix} \mathbf{Y}_d & \mathbf{0} \end{bmatrix} \begin{bmatrix} \Delta \mathbf{v}_d \\ \Delta \mathbf{v}_L \end{bmatrix} + \begin{bmatrix} \mathbf{W}_d \\ \mathbf{0} \end{bmatrix} \mathbf{x} =
$$

$$
\left\{ \begin{bmatrix} \mathbf{Y}_d & \mathbf{0} \end{bmatrix} \begin{bmatrix} \mathbf{Z}_{dd} & \mathbf{Z}_{dL} \\ \mathbf{Z}_{Ld} & \mathbf{Z}_{LL} - \mathbf{Z}_L \end{bmatrix} + \mathbf{I} \right\} \begin{bmatrix} \mathbf{W}_d \\ \mathbf{0} \end{bmatrix} \mathbf{x}
\tag{5}
$$

where

$$
\mathbf{Z}_{bus} = (\mathbf{Y}_{bus})^{-1} = \begin{bmatrix} \mathbf{Z}_{bus\,dd} & \mathbf{Z}_{bus\,dL} \\ \mathbf{Z}_{bus\,Ld} & \mathbf{Z}_{bus\,LL} \end{bmatrix}
$$

Let now the free system response of (3) be given by

$$
\mathbf{x}(t) = \sum_{j=1}^{n} (\alpha_j) \boldsymbol{\xi}_j e^{\lambda_j t}
$$

$$
\alpha_j = (\boldsymbol{\eta}_j^T) \mathbf{x}^o
\tag{6}
$$

where $\lambda_j (j = 1,...,n)$ are the system eigenvalues and $\boldsymbol{\eta}_j$, and $\boldsymbol{\xi}_j$ are the corresponding left and right eigenvectors; $\mathbf{x}_o$, is the initial state condition.

Inserting (6) into (4) yields

$$
\mathbf{Y}_{bus}^m \Delta \mathbf{V}(\lambda) = \mathbf{I}_{mod}(\lambda)
\tag{7}
$$

where

$$
\mathbf{Y}_{bus}^m = \begin{bmatrix} \mathbf{Y}_{dd} - \mathbf{Y}_d & \mathbf{Y}_{dL} \\ \mathbf{Y}_{Ld} & \mathbf{Y}_{LL} - \mathbf{Y}_L \end{bmatrix}
$$

and

$$\Delta \mathbf{I}_{mod}(\lambda) = \begin{bmatrix} \mathbf{W}_d \\ 0 \end{bmatrix} \mathbf{x} = \begin{bmatrix} \mathbf{W}_d \\ 0 \end{bmatrix} \sum_{j=1}^{n} (\mathbf{\eta}_j^T \mathbf{x}_o) \xi_j e^{\lambda_j t}$$

(8)

For a swing mode of interest, Eq. (7) allows modal voltages to be computed as a function of modal quantities. Because of its linearity, modal solutions can be associated with a single mode or mode combinations.

Once modal voltages are computed, the phase of the modal contributions is used to identify geographical zones showing coherent behavior. Buses experiencing the largest magnitude of oscillation indicate zones where voltage support is expected to have the largest influence on the inter-area mode of concern. These buses are then grouped together using the phase information to form a voltage control area.

Modal reactive power

In addition to voltage sensitivities, reactive power can also be used as a sensitive indicator of devices (machines and SVCs) having a large influence on the modes of interest. Explicit equations showing the dependence of reactive power deviations can be obtained from the above model noticing that

$$\Delta \mathbf{Q}_d = [\mathbf{I}_d^o]\Delta \mathbf{v}_d + [\mathbf{V}_d^o]\Delta \mathbf{i}_d$$

(9)

Where

$$\Delta \mathbf{Q}_d = \begin{bmatrix} \Delta Q_{d_1} & \Delta Q_{d_2} & \cdots & \Delta Q_{d_m} \end{bmatrix}^T$$

$$\Delta \mathbf{V}_d = \begin{bmatrix} \Delta V_{d_1} & \Delta V_{d_2} & \cdots & \Delta V_{d_m} \end{bmatrix}^T$$

$$\mathbf{I}_d^o = \begin{bmatrix} -I_{Q_1}^o & -I_{Q_1}^o & & \\ & & -I_{Q_1}^o & -I_{Q_1}^o & \\ & & & \ddots & \\ & & & & -I_{Q_1}^o & -I_{Q_1}^o \end{bmatrix}$$

$$\mathbf{V}_d^o = \begin{bmatrix} V_{Q_1}^o & -V_{D_1}^o & & \\ & & V_{Q_1}^o & -V_{D_1}^o & \\ & & & \ddots & \\ & & & & -V_{Q_1}^o & -V_{D_1}^o \end{bmatrix}$$

Solving for $\Delta \mathbf{v}_d$ and $\Delta \mathbf{i}_d$ from (4) and (5) in terms of $\mathbf{x}$ and substituting into (9) yields

$$\Delta \mathbf{Q}_d(\lambda) = \mathbf{M}_Q^o \mathbf{W}_d \sum_{j=1}^{n} \alpha_j \xi_j e^{\lambda_j t}$$

(10)

with

$$[\mathbf{M}_Q^o] = \left\{ [\mathbf{I}_d^o] \mathbf{Z}_{dd} + [\mathbf{V}_d^o] \mathbf{Y}_d \mathbf{Z}_{dd} + [\mathbf{V}_d^o] \right\} \mathbf{W}_d$$

Using the buses and reactive VAR sources that have large participation in the interarea modes of concern, voltage control area are formed and the most effective locations for voltage control are identified.

The preceding discussion provides a base for choosing transmission buses and reactive power sources having a strong influence on system behavior.

Viability Studies

Preliminary studies have been conducted to assess the ability of wide-area voltage support on the Mexican interconnected system. The study is based on a 6-area, 130-machine, 12 SVCs and 346-bus dynamic model of the MIS in which critical load areas are represented in detail.

Detailed modeling of SVCs including the control loop representation was used to determine dynamic characteristics.

Modal-based identification of critical voltage areas

Selection of suitable reactive power areas for voltage control is a crucial step in secondary regulation. A two-step hierarchical procedure for coordinating reactive power sources is developed and tested. The first step identifies buses and reactive power sources having a large participation in both the critical inter-area modes and system-wide voltage control. This enables to identify pilot buses and reactive areas showing a common behavior. In the second step, the available SVC voltage-VAR controls are coordinated with other reactive power sources to maintain adequate reactive power reserves for major system disturbances.

Figure 13 gives the normalized mode shape of the slowest inter-area mode in the system showing top participating areas. As can be seen in Figure 13, mode is highly observable in the north and south systems. While not directly shown in the diagram, machines with the largest participation are MTY U1-U5 in the Northeastern system, SLP in the Western system, and VDV in the Peninsular system.

Hierarchical control of static var compensators

Dynamic voltage support by SVCs at critical substations in the MIS is known to influence major inter-area modes [27]. Modal power flow analysis was used to understand the distribution of modal power across the system as well as to determine the most efficient SVC locations to control system voltage.

Examination of modal voltages in Figure 14 suggests that voltage control at SVCs GUE, ESA and XUL has the potential to impact system behavior.

Based on the study of modal information and all measurement sites available, the controllers were coordinated using the approach in [31].

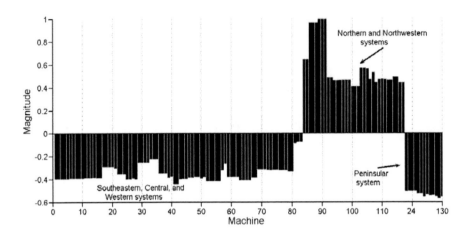

Figure 13. Normalized speed deviation of system generators for mode 1 showing the distribution of the mode

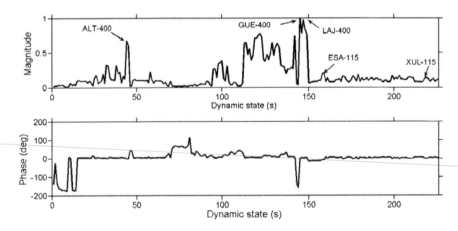

Figure 14. Normalized speed deviation of system generators for mode 1

Coordination of reactive power sources

The reactive power setpoint of major SVCs was optimized to ensure enough margin of reactive power. Using the approach outlined in Section [31,32] the overall reactive margin was optimized. In the results to be described below all SVCs are considered as candidates for VAR coordination. For clarity of illustration only results describing the impact of dominant SVCs are included.

Figure 15 shows modal voltages and reactive powers computed using (7) and for two operating scenarios:

a) Base case. SVCs in constant reference voltage control mode (primary voltage control)

b) Coordinated voltage-VAR control at various SVCs

Comparison of the modal voltages computed using (7) and (10) in Figure 15a for the base case condition with that with coordinated VAR control suggests that coordination of reactive power controllers increases the participation of machines close to the north-south interface and machines in the Peninsular system in the inter-area mode.

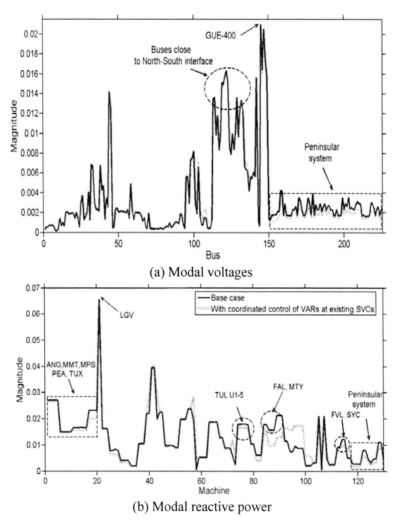

(a) Modal voltages

(b) Modal reactive power

Figure 15. Dominant bus voltage deviations. Outage of Laguna Verde unit # 1.

This is more clearly illustrated in Figure 15b that shows modal reactive power for the two scenarios above. It is seen that VAR coordination leads to a greater modal participation of machines in the Southeastern system (mainly ANG.MPS,PEA, and TUX) and machines in the Central (TUL), Northeastern (MTY) and Peninsular systems. Of significance, the large participation of the Laguna Verde (LGV) nuclear power station, the largest in the system, located in the southeastern network shows that this machine has a strong leverage on system dynamic performance.

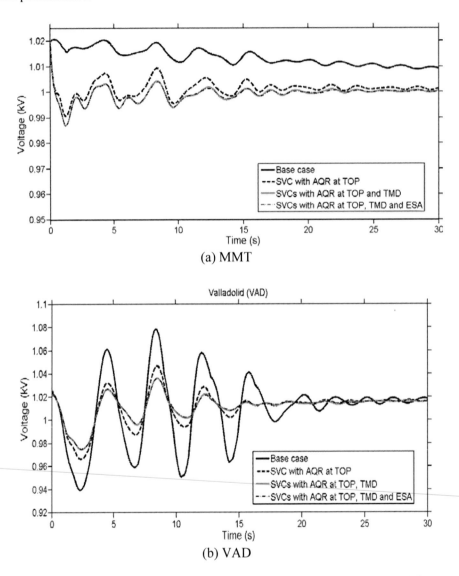

(a) MMT

(b) VAD

Figure 16. Comparison of 400-kV bus voltage magnitudes with and withour VAR control for outage of Laguna Verde unit # 1

Detailed small and large transient stability studies were conducted to validate the effectiveness of the coordinated wide-area dynamic reserve control concept. Selected dynamic simulations below were made using a large scale digital stability program are described below.

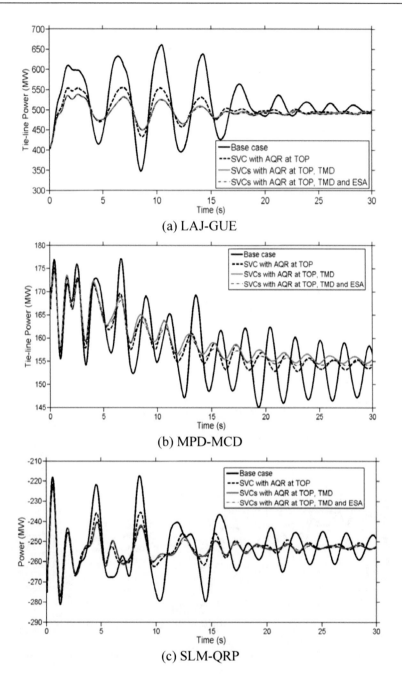

(a) LAJ-GUE

(b) MPD-MCD

(c) SLM-QRP

Figure 17. Comparison of tie-line power swings on major system interconnections following outage of Laguna Verde unit # 1

Effect on System Transient Behavior

Critical to voltage stability in the MIS is the loss of main generators in the system. Figures 16 and 17 show the system responses to the outage of unit # 1 of the Laguna Verde power station for various control alternatives. The SVCs participating in the secondary voltage control scheme are TOP, TMD, and ESA. Other SVCs were initially maintaining constant terminal voltage.

This contingency excites the 0.37 Hz mode and causes severe voltage and power swings along major interconnections. Examination of these results demonstrates that SVCs can have an important impact on the damping of post-disturbance power-angle and voltage swings due to the action of coordinated voltage control. Preliminary analyzes indicate that the results are insensitive to.

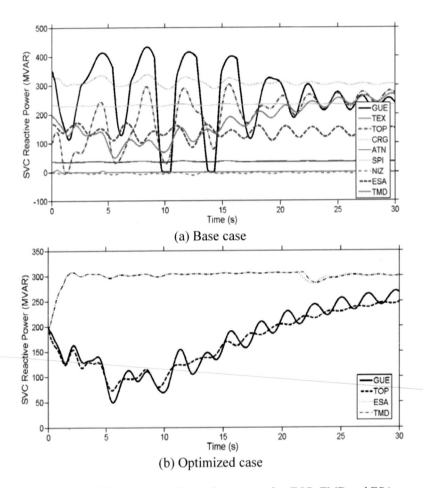

(a) Base case

(b) Optimized case

Figure 18. Reactive power. SVCs with secondary voltage control at TOP, TMD and ESA

Comparing Figures 18a with 18b shows that coordinated voltage control results in a better distribution of reactive power among the various SVCs. In the latter case, the SVCs operate in a more efficient and coordinate fashion to meet the operating target. By providing voltage control at the most effective system buses, damping can also be improved as is evident in Figs. 17 and 18.

This centralized approach improves coordination and ensures a better control of overall system dynamics.

SUMMARY AND CONCLUSIONS

This chapter discusses the experience with the application of advanced angle and voltage and control strategies to enhance power system angle and voltage stability. A methodology is presented for evaluating the benefits of optimizing the operation of several transmission reactive power resources. Experience with adaptive wide-area angle controls in the Mexican system shows that proper implementation of response-based special protection systems remedies the possibility of voltage collapse and uncontrolled islanding of the system.

The effectiveness of the improved remedial actions is dependent on several factors such as the timing of the control action and the amount, location and logic of generator (load) trip logic. These are aspects that warrant further investigation.

The studies also illustrate the potential for improved small and transient stability performance by means of static VAR compensators. By identifying transmission buses and reactive sources having the largest influence on a mode of concern, the best control approach can be identified and used to coordinate the operation of multiple VAR controllers. It is shown that system performance and security is improved by allowing greater reactive reserves for emergency conditions.

Despite these improvements, practical implementation of WAMS-based protection systems is still likely to be a challenging task. Numerical studies in this section provide some guidance in designing control functions to optimize their use and location. They also provide the basis for coordination with other protection schemes.

REFERENCES

[1] Anderson, PM; LeReverend, BK. "Industry experience with special protection systems IEEE/CIGRE Committee Report", *IEEE Trans. on Power Systems*, vol. 11, no. 3, August 1996, 1166-1179.

[2] Fouad, AA. (Chairman), "Dynamic security assessment practices in North America", Working Group on Dynamic Security Assessment, Power System Engineering Committee, *IEEE Trans. on Power Systems*, vol. 3, no. 3, August 1988, 1310-1321.

[3] Rovnyak, SM; Taylor, CW; Thorp, JS. "Performance index and classifier approaches to real-time, discrete-event control", *Control Engineering Practice*, vol. 5, no. 1, 91-99. 1997.

[4] Carson, W; Taylor, Dennis, C. Erickson, Kenneth E. Martin, Robert E. Wilson, Vaithianathan Venkatasubramanian, "WACS – Wide-Area Stability and Voltage Control System: R & D and online demonstration", *in Proc. of the IEEE*, vol. 93, no. 5. May 2005, 892-906.

[5] Miroslav Begovic, Damir Novosel, Daniel Karlsson, Charlie Henville, Gary Michel, "Wide-area protection and emergency control", *in Proc. of the IEEE*, vol. 93, no. 5, May 2005, 876-891.

[6] Lachs, WR. "Wide-area system protection scheme against extreme contingencies", *Proc. of the IEEE*, vol. 93, no. 5, May 2005, 1004-1027.

[7] Kevin Tomsovic, David E. Bakken, Vaithianathan Venkatasubramanian, Anjan Bose, "Designing the next generation of real-time control, communications, and computations

for large power systems", *in Proc. of the IEEE*, vol. 93, no. 5, May 2005, 876-891.

[8]   Aboytes, F; Arroyo, G. "Security assessment in the operation of longitudinal power systems", in *IEEE Trans. on Power Systems,* vol. PWRS-1, no. 2, May 1986, 225-232.

[9]   Castellanos, RB; Messina, AR. *Robust Stability and Performance Analysis of Large Scale Power Systems with Real Parametric Uncertainty*, New York: Nova Science Publishers, 2009.

[10]   Silva Peruyero, MA; Melendez Roman, CG. *Phasor measurement unit (PMU) applications in the transmission network of CFE,* Cigre Conference 2006, Paris France.

[11]   Enrique Martinez "SIMEFAS: A Phasor Measurement System for the Security and Integrity of Mexico's Electric Power System" *IEEE power & Energy Society 2008 General Meeting*, Pittsburgh, Pennsylvania USA 20-24 July 2008.

[12]   Phadke, AG. "The wide-world of wide-area measurement," *IEEE Power and Energy Magazine*, 52-64, Sept./Oct 2008.

[13]   Sada Gámiz, J; Aboytes, F. "Operation and control of the Mexican interconnected system", in *IEEE Trans. on Power Apparatus and Systems,* vol. *PAS-103, no.8,* August, 1984, 2081-2088.

[14]   Kosterev, DN; Esztergalyos, J; Stigers, CA. "Feasibility study of using synchronized phasor measurements for generator dropping controls in the Colstrip system", in *IEEE Trans. on Power Systems*, vol. 13, no. 3, August, 1998, 755-761.

[15]   Arturo Roman Messina (Editor),*Inter-Area Oscillations in Power Systems: A Nonlinear and Nonstationary Perspective*, New York: Springer Verlag, 2009.

[16]   Keith H; Holbert, Gerald T; Heydt, Hui, Ni. "Use of satellite technologies for power system measurements, command, and control", *in Proceedings of the IEEE*, vol. 93, no. 5, May 2005, 947-955.

[17]   Taylor, CW. *Power System Voltage Stability*, New York: McGraw-Hill, 1994.

[18]   Mariani, E; Murthy, SS. Control of Modern Integrated Power Systems, Advances in Industrial Control, Springer, London, 1997.

[19]   Kamwa, I; Béland, J; Trudel, G. "Wide-area monitoring and control at Hydro-Québec: Past, Present and Future", *2006 IEEE Power Engineering Society General Meeting,* Montreal, Quebec.

[20]   Lefebvre, H; Fragnier, D; Boussion, JY; Mallet, P; Bulot, M. *"Secondary coordinated voltage control system: feedback of EDF",* ", in Proc. of the 2000 Power Engineering Society Summer Meeting, 16-20 July 2000, Seattle, Washington, USA.

[21]   Taranto, GN; Martins, N; Falcao, DM; Martins, ACB; dos santos, MG. *"Benefits of applying secondary voltage control schemes to the Brazilian system"*, in Proc. of the 2000 Power Engineering Society Summer Meeting, 16-20 July 2000, Seattle, Washington, USA.

[22]   John, J. Paserba, *"Secondary voltage-var controls applied to static compensators (STATCOMs) for fast voltage control and long term var management"*, in Proc. of the 2002 Power Engineering Society Summer Meeting, 2002.

[23]   Claudio, A. Cañizares, Claudio Cavallo, Massimo Pozzi, Sandro Corsi, "Comparing secondary voltage regulation and shunt compensation for improving voltage stability and transfer capability in the Italian power system," *Electric Power Systems Research*, vol. 73, 67-76, 2005.

[24]   Sandro Corsi, "The coordinated automatic voltage control of the Italian transmission grid- Part I: Reasons of the choice and overview of the consolidated hierarchical

system", *IEEE Trans. on Power Systems*, vol. 19, no. 4, November 2004, 1723-1732.

[25] Martins, N. "The new CIGRE task force on coordinated voltage control in transmission networks", *in Proc. of the 2000 Power Engineering Society Summer Meeting*, 16-20 July 2000, Seattle, Washington, USA.

[26] Paul, JP; Leost, JY; Tesseron, JM. "Survey of the secondary voltage control in France: present realization and investigations", *IEEE Trans. on Power Systems*, vol. PWRS-2, no. 2, May 1987, 505-511.

[27] Roman Messina Messina, A. Miguel Angel Avila Rosales, "Analysis of inter-area damping enhancement by static VAR compensators in longitudinal power systems", *Control Engineering Practice*, vol. 5, no. 1, 1997, 117-122.

[28] Jesus González Flores, Cesar Fuentes Estrada, Miguel Angel Avila Rosales, "Mexican grid uses FACTs for greater flexibility", *Modern Power Systems*, June 1999, 33-35.

[29] Messina, AR; Hernandez, H; Barocio, E; Ochoa, M; Arroyo, J; "Coordinated application of FACTS controllers to damp out inter-area oscillations", *Electric Power Systems Research*, 2002, 43-53.

[30] Prabha Kundur, *Power System Stability and Control*, New York: McGraw-Hill, 1993

[31] Burchett, RC; Happ, HH; Vierath, DR; Wirgau, KA. "Developments in optimal power flow", *IEEE Trans. on Power Systems*, vol. PAS-101, no. 2, 406-414.

[32] *Optimization of reactive Volt-ampere (VAR) sources in system planning,* volume 1: Solution techniques, computing methods and results, EPRI EL-3729, Final Report, November 1984.

In: Electric Power Systems in Transition   ISBN: 978-1-61668-985-8
Editors: Olivia E. Robinson, pp. 171-195   © 2010 Nova Science Publishers, Inc.

Chapter 4

DESIGN AND APPLICATION OF A PROPOSED OVERCURRENT RELAY IN RADIAL DISTRIBUTION NETWORKS

A. Conde and E. Vazquez[*]

Universidad Autónoma de Nuevo León, Facultad de Ingeniería
Mecánica y Eléctrica, A.P. 36-F, CU, CP 66450, San Nicolás
de los Garza, Nuevo León, México

ABSTRACT

This chapter contains the design and application criteria for a proposed overcurrent relay. This relay uses independent functions to detect faults and to calculate the operation time. Two main functions are proposed, the first function offers a rather faster operation for low faults currents and greater sensitivity and the second function provides sensitivity for the detection of high impedance faults (HIF). The proposed relay have a negative sequence detector and positive sequence detector for confirm the presence of a fault, thus the negative sequence relay coordination is not necessary. The line protection scheme is simplified. The functional changes introduced simplify the setting process by the user with a minimal change in the relay's firmware and without change of relay hardware.

I. INTRODUCTION

The overcurrent relay is widely used in many protection applications throughout power systems. These devices provide fast operation at high current and slow operation at low current and, as the fault current is a function of the fault location the coordination with other overcurrent devices is possible. This behaviour is characteristic of overcurrent relays, and it has been demonstrated that it is appropriate for the protection of electrical systems in which operation above the nominal values is frequent but temporary.

However the growing load-ability of the electrical systems require the relay pickup settings to be increased, so that the need for greater sensitivity of the relays during an operational state of low demand is more significant. The short circuit current has a growth limited by the lack of investment in new generation power stations and/or by the reconfiguration of the electrical system in less meshed topologies with the intention of reducing the short circuit levels, avoiding the replacement of switches and other primary equipment. On the other hand, the load current is subject to the dynamics of industrial and commercial development, population growth and increase of electrical consumption.

Due to the above operation conditions, the application of time overcurrent relays in power systems has serious limitations in terms of sensitivity and high backup times for minimum fault currents. The high load current and different time curves for overcurrent protection devices, such as fuses and reclosers, reduce the reliability and security of the relay. The overcurrent coordination is carried out using maximum fault currents (3–5% of all faults) during maximum demand conditions (lasting only for a total of a few minutes per day) because of the convergence of overcurrent relay time curves for high fault currents; for other fault types and other current demand situations, the time curves diverge for minimum fault currents, and the backup times are much longer. This situation is not as convenient when it occurs in backup protection; due to the nature of the overcurrent relay, operating times are long, forcing the system to tolerate non-permissive currents, resulting in thermal and mechanical stress that could be minimized. The appearance of distributed generation (DG) and unconventional sources in low voltage networks may result in a change of the fault response [1,2]. The operation times of the overcurrent relays (primary and backup) can be rendered excessive by the topology diversity of the network.

The applications of overcurrent relays in distribution networks have been reported [3-8]. Various methods are proposed to solve the functional limitations of overcurrent relay. The use of relays based on differential phase to phase currents is proposed [4]. The information containing distorted current signals is used to offset the saturation effect of the CT [5]. An overcurrent directional relay based on ANN is proposed [6]. Symmetrical components are used to improve selectivity [7]. And at last, a methodology based on voltage overlapping signals to improve awareness of protection in radial distribution systems is presented [8].

An adaptive logic has previously been suggested at a substation level in which the relay settings could be modified from a central computer in the substation [9,10]. Updating adjustments of relays through communication channels using online algorithms to coordinate relays has also been proposed [11,12]. In isolated (rural) or highly connected networks in which it is not cost-effective to implement an adequate communication strategy, it is possible to carry out an automatic set-up relay using local current only. In this article, we propose an overcurrent relay with dynamic setting; this relay does not require communication channels to modify their setting parameters.

The proposed relay has two main functions, reduce operating times of primary and backup devices ensuring automatic coordination and provide sensitivity for the detection of high impedance faults (HIF).

The pickup current of conventional overcurrent relays has two different functions, the detection of the fault condition and the determination of time of operation. In the current electrical systems that present unequal increases in load levels and the short circuit ability, the

* Corresponding author: (con_de@yahoo.com, evazquez@gama.fime.uanl.mx).

task of discriminating between a condition of overload and a fault is hardly obtained with a single setting. We propose to separate the functions of fault detection and to calculate the operation time in specific functions. The principles reported are applied for radial systems with one supply source or more than one supply source (directional relay). The proposed relay does not represent a global solution to the problem of overcurrent relay application but nevertheless it offers functional advantages that allow obtaining another functional approach of the protection.

II. INVERSE TIME OVERCURRENT DIGITAL RELAY

The basic model and digital implementation of an overcurrent relay system is presented in [13]. In this section, we present the functional structure as the basis of the proposed relay (Figure 1). The input signals are the fundamental current phasor I_k^r and the pickup current I_{pickup}. The relay generates the no lineal function $H(I_k)$, where $I_k = I_k^r / I_{pickup}$ is the operating current. The function $H(I_k)$ is integrated, and the output integrator signal is:

$$G_k = \Delta t \sum_{k=1} H(I_k)$$

(1)

where G_k is the accumulated value of the integrator in the sample k and Δt is the sampled period.

The operating condition is obtained when:

$$\Delta t \sum_{k=1}^{k_{op}} H(I_k) = J(I_k)$$

(2)

The relay operation is complete when $k = k_{op}$ and equation (2) is satisfied. The functional relationship for overcurrent relays is obtained from $T = k_{op} \Delta t$ and equation (2). For constant fault current:

$$T(I) = \frac{J(I)}{H(I)}$$

(3)

Then, in each sample period the ratio remains:

$$T(I_k) = \frac{J(I_k)}{H(I_k)}$$

(4)

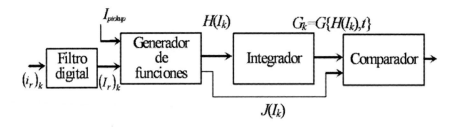

Figure 1. Functional diagram of an inverse time phase overcurrent relay

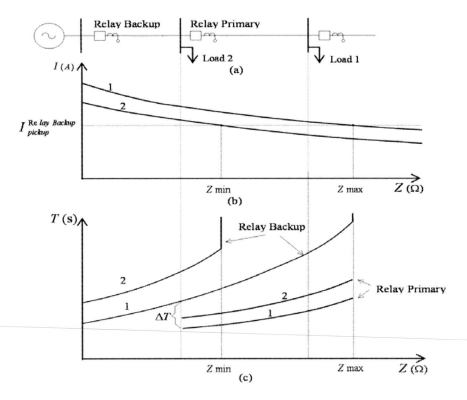

Figure 2. Effect of short-circuit current in the reach of time overcurrent relay.

For variable fault currents and $J(I_k) = K$ using equations (4) and (2), we obtain:

$$\sum_{k=1}^{k_{op}} \left(\frac{1}{T(I_k)} \right) \Delta t = 1 \tag{5}$$

In equations (3 and 4), we observe that the functional relationship between $J(I_k)$ and $H(I_k)$ defines the characteristics of overcurrent relays. The shape of the time curve produced is dependent on the $H(I_k)$ function when $J(I_k) = K$. We can modify this function to obtain different time curves for enhanced coordination.

III. OPERATING LIMITS OF THE OVERCURRENT RELAY

The overcurrent protection uses current as the only indicator of fault location. However, the fault current depends on the fault type and the pre-fault steady state. Moreover, the maximum load current can be similar in magnitude to the minimum fault current; this makes it difficult to correctly discriminate between the normal stable state and the fault condition.

The limitations of an overcurrent relay are illustrated in the radial grid in Figure 2a. Relay Backup is used as a reference point that provides primary protection of the line itself and backup of the adjacent line. Figure 2b shows a graph of the variation in the short-circuit current as a function of the electrical distance Z to evaluate the sensitivity in the protection. Figure 2c shows the time-electrical distance $T=f(Z)$ used to evaluate the operation time. In Figure 2b, curve 1 corresponds to the maximum level of the short-circuit current (three-phase fault at maximum demand), and curve 2 corresponds to the minimum level (two-phase fault at minimum demand). The value of the adjustment current I_{pickup} corresponds to Relay Backup.

One limitation of the overcurrent protection is that its reach (the length of the protection zone) depends on the type of short-circuit and the system operation. Figure 2b shows that the reach of Relay Backup moves between the limits Z_{min} and Z_{max}, and can stop providing proper backup to the adjacent line for minimum generation. Because of these factors, the reach of the overcurrent relay changes dynamically, depending on the operational state of the electrical grid; the protection could be lost in minimum operating conditions. This is particularly the case for the phase protection, in which the maximum load current defines the pickup current of the relay. Consequently, the sensitivity limitation of overcurrent relays is poor fault detection in minimum power generation conditions.

Another problem in overcurrent protection is the high backup time for minimum fault currents (Figure 2c). This behavior is characteristic of overcurrent relays demonstrated to be appropriate for the protection of electrical systems in which operation above the nominal values is frequent and temporary. This situation is not as convenient when it occurs in backup protection; due to the nature of the overcurrent relay, operating times are long, forcing the system to tolerate non-permissive currents and resulting in thermal and mechanical stress that could be avoided. The load current (high pickup setting) and the divergence of the relay's time curve for poor fault currents result in high backup times (see curve 2 for Relay Primary and backup Relay Backup in Figure 2c). When both the primary and the backup overcurrent protection have different time curves, adequate time coordination is difficult. Therefore, the time limitation of overcurrent relays is high backup times for both minimum fault currents and different time-curve devices.

IV. APPLICATION CRITERIA OF PROPOSED RELAY IN POWER SYSTEMS

The proposed relay has two different functions depending on the magnitude of the fault current (Figure 3), the first is an operation designed to improve the times of the relay for fault currents greater than the sensitivity downstream protective device, and increase its sensitivity in comparison of similar conventional. It requires a correct discrimination between fault conditions and normal operation of the system as power transfer and variation of load. Fault detectors measure the negative sequence current and positive sequence current, so the detection function with negative sequence relay [14] is not necessary in the line protection scheme because it is included in the proposed relay, we have the advantage of not coordination is needed.

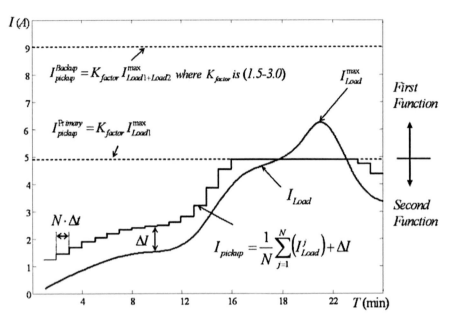

Figure 3. Proposed relay functions.

The second function is to detect faults with less than current values downstream adjustment device, this function represents a backup for ground relays. The pickup current I_{pickup} for the proposed overcurrent relay will be a function of the load current I_{Load} (Figure 3). Where ΔI represent a safety margin, with a proposed value of 15% of the maximum load current, Δt is the sample period and N must be selected in such a way that the interval $N \cdot \Delta t$ lasts between one and several minutes as the integration time used in demand measures. Previously, functions that implied an increase in the computational resource represented a disadvantage; at the present time the relays have the hardware to make the measurement functions, and also to execute the proposed functions. This pickup function provides increased sensitivity, because the value of I_{pickup} is also small due to minimal demand conditions.

V. FIRST FUNCTION

A. Time Coordination

The basic idea for time coordination is to satisfy equation (6) [3] for any current value.

$$T_{backup} = T_{primary}\left(I_k^{primary}\right) + CTI$$

(6)

where T_{backup} is the time curve of the backup relay, $T_{primary}\left(I_k^{primary}\right)$ is the time curve of the primary overcurrent device, $I_k^{primary}$ is the operating current of the primary device for each sample k, and CTI is the coordinating time interval (0.2-0.4 s). The operating current of the primary relay is calculated using the pickup current of the primary device and the fault current $I_k^{primary} = I_k^{system} / I_{pickup}^{primary}$.

The main purpose is to find a time element function T_{backup} that ensures that the backup relay operates with a constant time delay relative to the primary device, for any fault current. For this to happen, it is necessary that the operation time of the backup can be determined from the time curve of the primary.

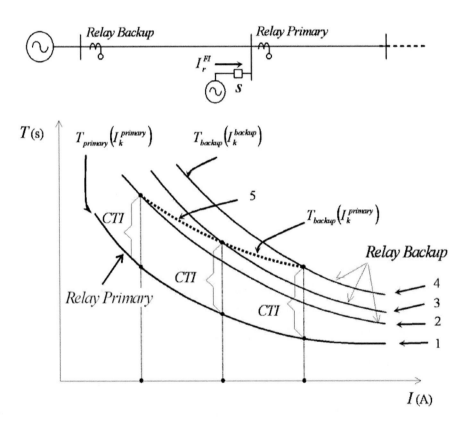

Figure 4. Proposed time curve of overcurrent relay

Figure 4 shows the overcurrent relay coordination system. The coordination was done using a computer simulation. By raising the load current (pickup setting), the backup time is increased, although both relays have the same time curve. To obtain the same backup time delay (CTI) for all fault currents, two different mechanisms are possible: the first is to change the dial time for each fault current (curves 2, 3 and 4 in Figure 4); and the second — a better solution — is to change curve 5. Curve 5 is not obtained using a dial time setting of the primary relay (curve 1) due to the load current. In order to change the overcurrent relay time curve, curve 5 needs to change its shape.

In Figure 4, we observed that curve 5 is similar to curve 1. For this to occur, it is necessary to use the pickup setting of the primary device to calculate the operating current. Then the proposed relay emulates the dynamics of the primary device to obtain a fast backup operation time. A minimum time curve for the backup device is obtained, because this is asymptotic to the primary pickup current. By comparing the conventional relay with the proposed relay, the first follows curve 4 whilst the second follows curve 5 of Figure 4. A reduction of backup time is obtained with the proposed relay. On the basis of these results, we considered the pickup current of the backup relay to be only a fault detector and not be used to calculate the operation time of relay.

The equation describing the proposed relay is obtained. The operating current used is the one in the primary relay (eq. 1):

$$G_k = \Delta t \sum_{k=1} H\left(I_k^{primary}\right)$$

$$where: \quad H\left(I_k^{primary}\right) = \frac{1}{T_{primary}\left(I_k^{primary}\right) + CTI} \tag{7}$$

The operating condition [16] is obtained when:

$$G_k = \Delta t \sum_{k=1}^{k_{op}} H\left(I_k^{primary}\right) = 1 \tag{8}$$

The relay operation is complete when $k = k_{op}$ and equation (8) is satisfied.

The infeed current effect in the overcurrent coordination is shown in Figure 4. Consider the situation when switch S closed. The infeed current I_r^{FI} accelerates the operation of the *Relay Primary*, though the backup time is the same (equal to the time of *Relay Backup*) and the *CTI* is bigger. With the coordination proposed the *CTI* is the same (Curve 5 equal to curve 1 in Figure 4). Under these circumstances the proposed backup relay is faster than a conventional backup relay.

The off-line computed time curve proposed is calculated using equation (7). If the time curve of the primary overcurrent device is analytical (digital relays), the setting curve is computed to directly substitute for the function $T_{primary}\left(I_k^{primary}\right)$. When the characteristic is not available (for example in fuses, electromechanical relays and reclosers), it is possible to calculate analytical expressions using fitting curve algorithms [17].

In Figure 4, we observed that curve 5 is similar to curve 1. For this to occur, it is necessary to use the pickup setting of the primary device to calculate the operating current. This results in a minimum time curve for the backup device as the backup curve is asymptotic to the pickup primary current (Figure 5). The shaded area is the increase in the sensitivity of the proposed relay compared to a conventional relay. The analytical time curves were analysed using IEC Standard 255-4 [**18**].

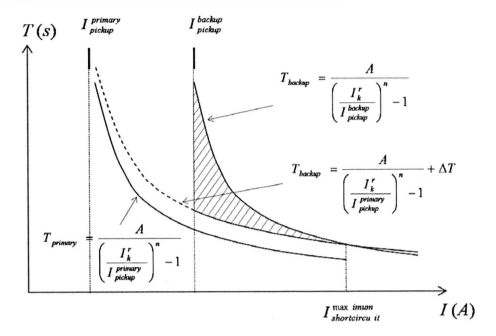

Figure 5. Proposed time curve

B. Coordination

The coordination example was carried out in the typical 13.8 kV distribution systems shown in Figure 6. It is not necessary to consider a more complex power system configuration, as the use of a complex power system does not reach an unexpected place. Most scenarios have the same effect on the operating current so the time overcurrent relay coordination process is carried out using pairs of relays. The coordination example is demonstrated in the radial lines with the assistance of the commercial software Aspen Oneliner. We observed that the backup time of *Relay B* (section *a–b*) is greater than that of the *proposed Relay B*. Therefore, the coordination proposed allows a rapid time curve to be selected for *Relay A*. The coordination between the *proposed Relay B* and *Relay C* is carried out in the same relay. Using the time curve (see equation (7)), coordination is automatic; even when there is an increase in the maximum fault current (topology changes or additional power generation), coordination is carried out and setting changes are not necessary.

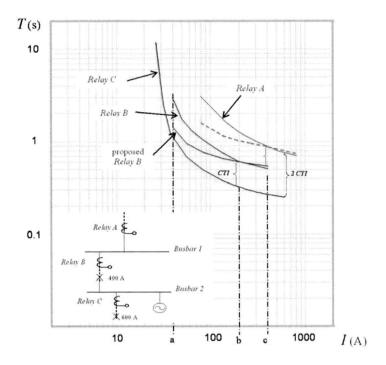

Figure 6. Time coordination example of overcurrent relays.

The coordination between the fuse, the proposed relay (B) and the conventional relay (A) is shown in Figure 7. The proposed relay curve is the same (plus CTI) as that of the maximum clearing time fuse curve. The coordination process between the conventional relay and the fuse can be achieved with 2CTI as a coordination interval or directly with the proposed time curve.

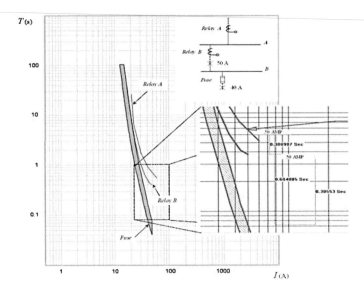

Figure 7. Time coordination example of fuse and overcurrent relay

In Figure 8, the coordination of a recloser and relay is shown using 13.8 kV radial systems. The proposed coordination is achieved with minimal backup time.

In the coordination test shown, we have observed that the minimum backup time is obtained. In addition, the coordination process occurs with the relay. Following this, coordination between the proposed relay and the overcurrent protection device (such as an electromechanical relay, fuse or recloser) is automatically obtained. The data necessary for coordination of the proposed relay are the data for the voltage system and impedance line. For data protection, the time curve and pickup of the primary device are needed. With this information available, coordination is achieved.

C. Test Operation in Steady State and Dynamic State

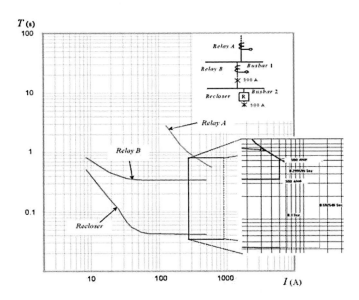

Figure 8. Time coordination example for the recloser and overcurrent relay

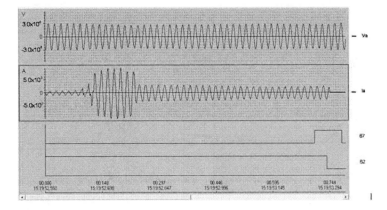

Figure 9. Fault signal recorded in a 34.5-kV distribution network

A fault log was used to test the signal file. This two-phase fault was logged in a 34.5-kV distribution grid. Figure 9 shows the record of the phase relay event and Table I shows the relay setting and its operating time. For the sake of simplicity, the values shown are the primary system. The maximum load current of the primary relay is 190 A and the maximum load current of the backup relay is 286 A. The operating time of a conventional relay is calculated by equation 5 and verified by application (Figure 9). If the proposed time coordination is used, the operating time is reduced by 0.2745 s. This is a quantitative example of the benefit of the proposed coordination.

The time coordination between *Relay B, Relay A* and the proposed *Relay A** (Figure 10) was evaluated in a dynamic fault current situation. The fault current signal and the integration process of the overcurrent relays with variable fault currents were obtained in both laboratory tests and by digital simulation. The proposed relay was developed with the assistance of the commercial software Labview and executed in a real-time DSP card. The output of the integrator was recorded. The dynamic fault current ($I_{shortcircuit}$) and the integrated value in *Relay B* ($G_k^{primary}$), *Relay A* (G_k^{backup}) and the proposed *Relay A** ($G_k^{backup*}$) are shown in Figure 11. For all relays, the time curves are inverse type [15]. We observed that the time interval between *Relay B* and *Relay A* is 0.61 s, although the operation time difference between *Relay A** and *Relay B* is 0.3 s (*CTI*). This highlights the advantage of the proposed time relay versus a conventional relay in backup zones.

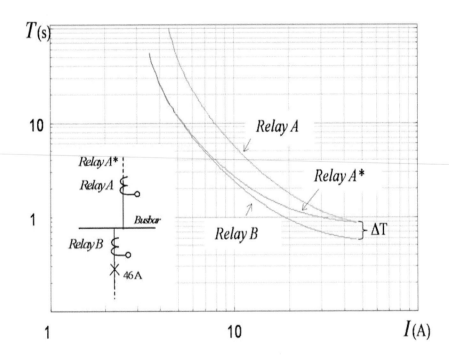

Figure 10. Time coordination in laboratory test

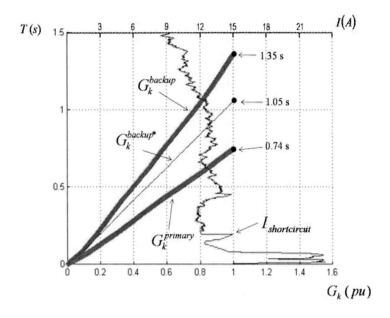

Figure 11. Accumulated value of the relays integrators in laboratory test

D. Fault Confirmation Logic

In situations in which a large load would be added on a feeder, the relay has detection fault logic (Figure 12) in order to supervise I_{pickup}. This logic includes a negative sequence verification and a high-level current detector, both combined in *OR* logic. The negative sequence was proposed to detect phase-to-phase faults and its setting has been previously described [**14**]. In low-voltage networks, the negative sequence current of phase-to-phase faults is higher than the maximum unbalanced negative sequence current, leading to the ability to obtain a good setting. The high-level detector is proposed to detect three-phase faults; this uses the same setting as a conventional time overcurrent relay. Therefore, this logic discriminates between a large load and a fault (symmetrical or asymmetrical).

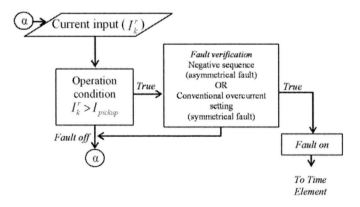

Figure 12. Control logic of a proposed pickup current

Table I. Overcurrent relay setting.

Conventional	Proposed
I_{pickup}=430 A	I_{pickup}=275 A (primary pickup)
Moderate inverse curve [15], Dial=0.2	
Time=0.7067 s or 42.4 cycles (f=60hz).	Time=0.4322 s or 25.9 cycles (f=60hz)

The fault detector performance was evaluated in the network shows in Figure 13a. The simulated operating sequence (PSCAD®) consists of three stages: unbalance in normal operating conditions (0.0 to 1.0 s), two-phase fault in adjacent line (1.0 to 1.5 s) and tripping of the adjacent line (power transfer, 1.5 to 2.0s). The performance of the relay A is evaluated; this relay must tolerate the unbalance condition, backup the fault in adjacent line, and not trip for power transference. During the state of imbalance in stable state (25% according to [3]) the algorithm produces no output, tolerating this condition; a value of 80 A [14] was used for this simulation (Figure 13b). During the fault, there is an appreciable value of negative sequence current I_2 seen by *Relay A*, allowing effective detection.

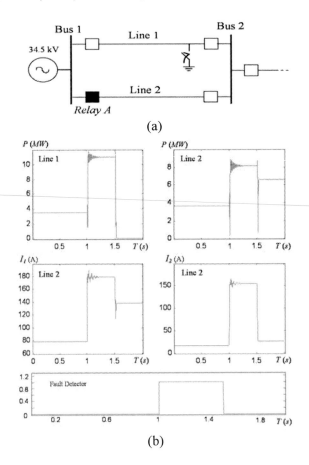

(a)

(b)

Figure 13. Evaluation of proposed fault detector.

In the last simulated sequence, tripping line 1 will trigger a power transfer I_1 in *Relay A*. This condition should be tolerated by the relay, allowing the load feed. The adjustment of the symmetrical fault detector was similar to that of a conventional overcurrent relay [16]. During this condition, the output of the proposed fault detector is not present; performance is satisfactory.

In Figure 13, the relay A must provide backup for two relays (this is the protection of multi-terminal lines), the time curve equation of relay A is defined by the slowest primary protection. Three-phase faults are detected by the high-level detector; this uses the same setting as a conventional time overcurrent relay. Thus the pickup of proposed relay never has greater than conventional relay.

E. Application Criteria for the First Function

Then the proposed relay uses the pickup current only as the fault detector; the operation time is calculated with the pickup current and with the time curve of the primary device. Therefore, the proposed relay emulates the dynamics of the primary device to obtain a fast backup operation time. The operation of overcurrent relays is good for maximum fault contributions, then the functional disadvantages occur during low demand conditions, thus the analysis is carried out emphasizing the advantages of proposed relay in the region of time curve near to pickup current.

The proposed relay can be implemented with the pickup current of the primary relay. In this way, both an increase of sensitivity and minimum backup time are obtained. The application presented is for radial systems with one or multiple supply sources because the pickup current of the backup relay could be greater or equal to the primary relay; thus the proposed relay can be applied without difficulty. Nevertheless the coordination in a ring-fed distribution network will have to be evaluated because in any given coordination pair the pickup current of the backup relay could be smaller than the pickup current of the primary relay. In these circumstances the proposed relay can be considered as an acceleration element of backup time of an actual relay as shown in Figure 14.

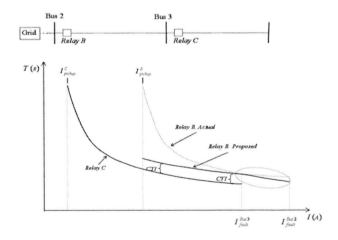

Figure 14. Proposed relay considered as an acceleration element of the backup time of the actual relay.

We observed that the time curve of the actual relay is without change, the proposed relay is addition in logic OR; any relay can send the outlet signal to the circuit breaker. Following this a combined curve is obtained that has the advantages of both relay, the actual relay with its convergent characteristic for three-fault ($I_{fault}^{Bus\ 3}$ to $I_{fault}^{Bus\ 2}$), and the proposed relay with fast backup time (I_{pickup}^{B} to $I_{fault}^{Bus\ 3}$). Fault confirmation logic for the proposed relay is not required.

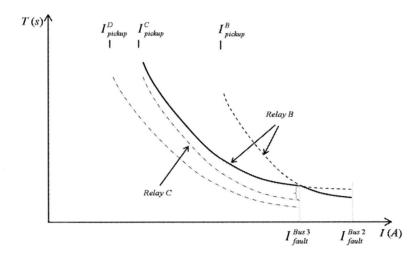

Figure 15. Proposed relay to increase the sensitivity and to reduce the backup time

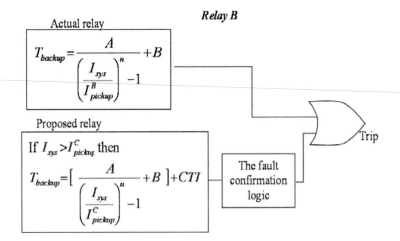

Figure 16. Application logic of the proposed relay

The proposed relay used in radial systems with one or multiple supply sources, increase the sensitivity of the protection and to reduce the backup time as shown in Fig 15. The power system is shown in Figure 14. The time curve of the proposed relay is asymptotic to the pickup current of the primary relay (*Relay C*). In this way the pickup of the proposed relay is activated when $I_{sys} > I_{pickup}^{C}$, after the output of the fault confirmation logic has verified whether to enable or disable the relay operation. The application logic is as Figure 16.

The coordination of the proposed relays compared with the actual is shown in Figure 17 for the power system shown in Figure 14. The coordination process of the actual relays will be conventional. The time curve of the proposed relay is as follows: the time curve equation of the proposed relay B is same than actual relay C plus *CTI* (eq. 7), assuming that relay C is slower than any other relay in the same bus. The same process is carried out for the coordination of relays A and B, with the time curve equation of proposed relay A the same as the actual relay B plus *CTI*. This process is carried out using the algorithm of the proposed relay, but the user will need to select the response curve of the primary protection device.

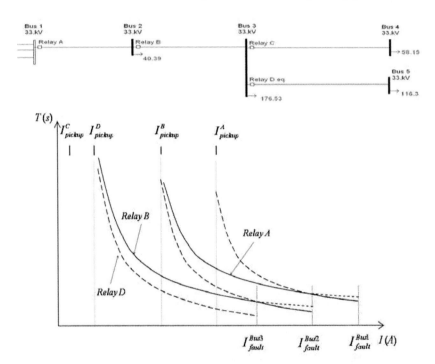

Figure 17. Coordination of actual and proposed overcurrent relays.

This application is evaluated for de use of in relay B in the feeder circuit of Figure 14; the resulting coordination is displayed in Figure 18.

The high benefit of proposed relay is in main feeder protection, where an adequate sensitivity of the overcurrent relay is difficult to obtain. If the minimum fault current in the remote bus of any feeder line is greater than the pickup current of the main relay the backup function is obtained. The example of coordination with approach two is showed in Figure 18.

The proposed relay (firmware) can be added to the actual relay, so the setting and coordination process of the actual relays are unchanged.

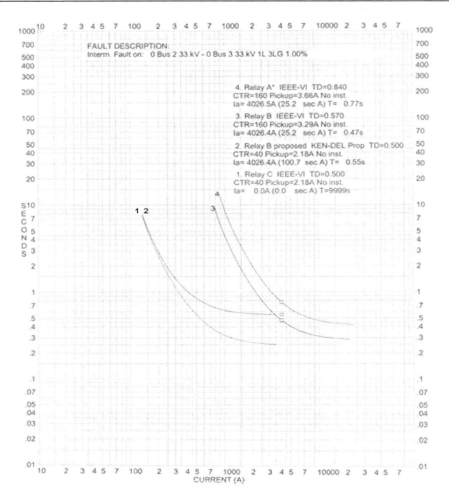

Figure 18. Coordination example of actual and proposed relay

VI. SECOND FUNCTION

The maximum load current is present only few minutes per day; the rest of the time, the load current has less value. Then the pickup current is advisable to define it based on the value of demand of the load current (Figure 3).

This adaptability operation can be represented through a characteristic represented in a time-impedance plane $T=F(Z,I_{pickup})$, where Z is electrical distance in ohms from the relay to the fault position; this characteristic, contrary to the conventional relay case, has a third variable, the pickup current I_{pickup}, and therefore it must be represented in a tri-dimensional form.

In Figure 19 we represent the characteristics $T=F(Z)$ for the conventional relay; the representation is tri-dimensional, as a comparison base for the adaptive case. In reality, the characteristic is in the corresponding plane to $I_{pickup}=50$ A for this example. The shaded region represents all the possible characteristics $T=F(Z)$ in different states of the system.

In Figure 20 we show the characteristics $T=F(Z, I_{pickup})$ of the adaptive relay. We observe that the operating zone of the relay has been into a surface with the adaptive concept. We generate two surfaces, 1 for two-phase faults and 2 for three-phase faults, which are limited by the variation interval of the pickup current I_{pickup} of the relay ($I_{pickup\ máx}= 50A$, $I_{pickup\ mín}= 25A$). We observe that in minimum load current we have a greater sensitivity and less operation time of the relay for the different power conditions.

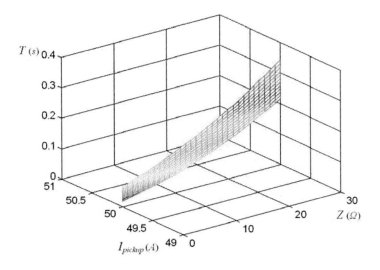

Figure 19. Conventional relay characteristic $T=f(Z)$

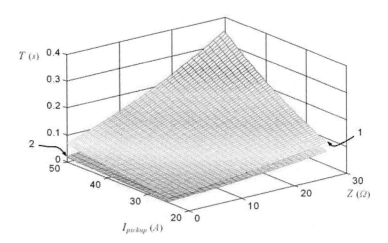

Figure 20. Adaptive relay characteristic $T=f(Z,I_a)$

The control logic of the pickup current (Figure 21) has the task of maintaining constancy of I_{pickup} during a fault to avoid pickup setting changes. If the line is de-energized ($I_k^r < \varepsilon$), the control logic assigns a maximum value '$I_{pumáx}$', which can be the setting of a conventional relay (dead line logic). The output signal T is done if a fault has been detected; the signal F is a blocking order. During a complete demand interval, the value I_{pickup} is fixed in the relay at the end of the previous interval. The action of low-pass filtering that is inherent in the demand concept simplifies the logic of the relay.

a. Overcurrent Relays

In this section, results from the three types of phase overcurrent relays are collated: conventional, negative sequence and adaptive. The sensitivity is also analysed.

Sensitivity analysis was carried out in the radial power system shown in Figure 22a. It is not necessary to consider a more complex configuration of the power system as mentioned in second function section. The minimum fault current is simulated in Bus 4. The variable load impedances simulate a multi-tap line (Z_3) and a variable load (Z_4). For analysis, the phase overcurrent relays are located in Bus 2 (Relay B). Sensitivity is determined by the following relationship:

$$\text{Sensitivity} = \frac{I_{minimum\ short\ circuit}}{I_{pickup}} \tag{9}$$

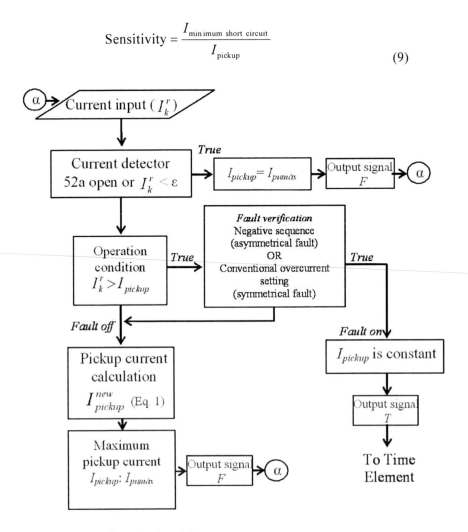

Figure 21. Control logic of an adaptive pickup current

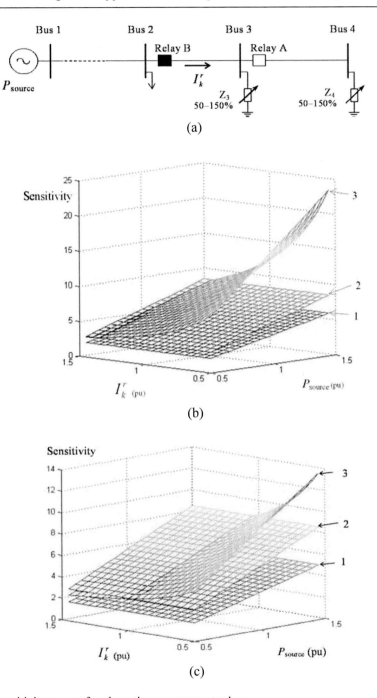

Figure 22. Sensitivity curves for phase time overcurrent relays

The different sensitivities expected for each relay pickup are calculated in different ways. For negative sequence relays and conventional relays, the pickup current method has been described previously [14,16]. For adaptive relays, although the limit of sensitivity is similar to that of the pickup of the primary device, in this sensitivity analysis we assume that adaptive relay is not limited maximum sensitivity.

The minimum short-circuit current is the phase-to-phase fault in Bus 4. The accepted value is 1.5 [16]. Figure 22b shows the relay sensitivities for different load currents (I^r_k) and source contributions (P_{source}). The proposed relay (plane 3), rather than the conventional (plane 1) and negative sequence (plane 2) relays, increases sensitivity for the minimum load. The conventional and negative sequence relays have a pickup setting that is independent of the dynamic load; this is observed in planes 1 and 2 in Figure 22b. Proposed relays have a setting that is dependent on the dynamic load; this characteristic result in plane 3, in which the adaptive relay has greater sensitivity compared with the other two relays in poor load currents. Under high power conditions (high P_{source}), the sensitivity of the three relays is higher (larger short-circuit currents), and the effect in the proposed relay is to increase sensitivity more than in the others relays. It can be observed that adaptive relays are more dependent than the other relays on the power contribution and load current for sensitivity.

In multi-terminal lines (Figure 22c), the negative sequence relay has the same or higher sensitivity than the adaptive and conventional relays; negative sequence relays are not affected by the load current, allowing easy current coordination with the high pickup setting for primary protection of multi-tap lines. Under maximum demand conditions, the sensitivity of negative sequence relays is higher than for other relays; however, under minimal demand conditions, the sensitivity of adaptive relays is higher. These conditions may change because of topology dependence or protective schemes, but the results of this study indicate the general trend.

Conventional relays have poor sensitivity, and negative sequence relays are best used in multi-terminal lines (but only for unbalanced faults). Proposed relays have a high sensitivity for radial and multi-terminal lines and can be used effectively for three-phase faults and phase-to-phase faults.

b. Test

The adaptive pickup current logic performance was tested with digital simulation. The test was done on the basis of the displayed power system in the Figure 22a. Figure 23 shows a typical protective system operation in a subtransmission and distribution electric network. The adaptive logic remains in normal conditions (load current) until the current is higher than in the pickup current (fault current). As the algorithm uses the demand current information, the setting of the pickup current (I_{pickup}) is constant in each demand interval. The operation logic of the adaptive pickup current is satisfactory.

The accumulated value of the integrator (G_k) or, analogically, the position of the induction disk in an electromechanical relay at any time, depends on I^r_k. In the operation, the integration of the function $T(I_k)$ is positive, increasing the accumulated value in the integrator (distance traveled by the disk towards the tripping position); on the other hand, in the reset zone, the integration is negative and decreases the accumulated value in the integrator (return of the disk towards the actuating position).

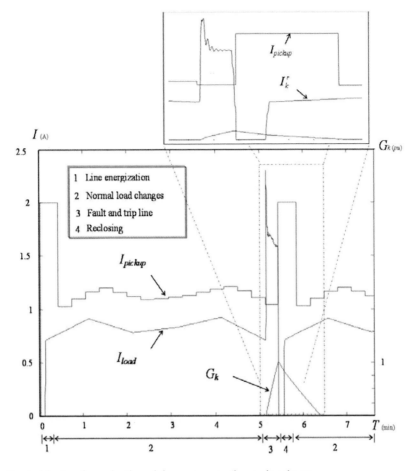

Figure 23. Control logic of an adaptive pickup current using a signal test

Figure 23 shows that the magnitude of the fault current is similar to the relay setting, then the fault can not be detected; but the second function proposed allows detection of the fault. High impedance faults are grounded, ground relays are enabled to detect these faults. However situations like high unbalance in the electrical grid causing high settings, then the ground protection may be insensitive. In addition to this, the impedance of the fault to the substation can be large, also may be common in some circuits that ground wire is lost through vandalism or connections sulphated, under these conditions the ground protection will have little chance of success. Under these conditions the second proposed function allows a backup to the ground relays.

Setting considerations of the second function proposed are highly dependent on the application, some examples are: 1) enables the defined time relay, should have no commitment to coordination, the setting of time is determined with or without delay. The choice may be established as a criterion of fit for the user; 2) time coordination between ground relay and proposed relay.

VII. CONCLUSION

Increased speed of operation and increased sensitivity of overcurrent relays can be obtained with two proposed functions; the first function offers a rather faster operation for low faults currents and greater sensitivity. The coordination process with actual relays is simplified. The second function provides sensitivity for the detection of high impedance faults (HIF) being a backup function protection to earth relays. For the protection functions, we propose to separate the functions of fault detection (pickup current) and the relay operation time (time curve). The functional changes introduced simplify the setting process by the user with a minimal change in the relay's firmware and without change of relay hardware.

REFERENCES

[1]   Girgis, A; Brahma, S, "Effect of distributed generation on protective device coordination in distribution system," *Power Engineering*, 2001. LESCOPE '01. 2001, 11-13 July 2001 Page(s):115-119.

[2]   So, CW; Li, KK. "Protection relay coordination on ring-fed distribution network with distributed generations," TENCON '02. Proceedings. 2002 IEEE Region 10 *Conference on Computers, Communications, Control and Power Engineering*, Volume 3, 28-31 Oct. 2002 Page(s):1885 - 1888.

[3]   ANSI/IEEE Std 141-1986, *IEEE Recommended Practice for Electric Power Distribution for Industrial Plants.*

[4]   Chen Yanxia, Yin Xianggen, Zhang Zhe, Chen Deshu, "The research of the overcurrent relays based on phase-to-phase differential current -adaptive setting and coordination," Transmission and Distribution Conference and Exposition, 2003 *IEEE PES,* Volume 1, 7-12 Sept. 2003 Page(s):250-255.

[5]   Tunyagul, T; Crossley, P; Gale, P; Zhao, J. "Design of a protection relay for use with a measuring CT," *Power Engineering Society Summer Meeting,* 2000. *IEEE*, Volume 3, 16-20 July 2000 Page(s):1390-1395.

[6]   Vishwakarma, DN; Moravej, Z. "ANN based directional overcurrent relay," Transmission and Distribution Conference and Exposition, 2001 *IEEE/PES*, Volume 1, 28 Oct.-2 Nov. 2001 Page(s), 59-64.

[7]   Lotfi-fard, S; Faiz, J; Iravani, R. "Improved Overcurrent Protection Using Symmetrical Components," *IEEE Transactions on Power Delivery,* Volume 22, Issue 2, April 2007 Page(s), 843-850.

[8]   Zamora, I; Mazon, AJ; Sagastabeitia, KJ; Zamora, JJ. "New Method for Detecting Low Current Faults in Electrical Distribution Systems," *IEEE Transactions on Power Delivery,* Volume 22, Issue 4, Oct. 2007 Page(s), 2072-2079.

[9]   Shah, KR; Detjen, ED; Phadke, AG. "Feasibility of adaptive distribution protection system using computer overcurrent relaying concept," IEEE *Transactions on Industry Applications*, vol. 24, No. 5, September/October 1988, 792-797.

[10]   Sachdev, MS; Sidhu, TS; Chattopadhyay, B. et. al., *"Design and evaluation of an adaptive protection system for a distribution network,"* Cigré Paper 34-202, París,

1995.

[11] Askarian, H. et al. "A new optimal approach for coordination of overcurrent relays in interconnected power systems," IEEE *Transactions on power delivery*, vol. 18, No. 2, April 2003.

[12] Abdelaziz, AY; Talaat, HEA; Nosseir, AI; Ammar, A. Hajjar, "An adaptive protection scheme for optimal coordination of overcurrent relays;" *Electric Power Systems Research*, Vol. 61, Issue 1, 28 February 2002, 1-9.

[13] Benmouyal, G. "Some aspects of the digital implementation of protection time functions," *IEEE Transactions on Power Delivery*, Vol. 5, No. 4, November 1990, 1705-1713.

[14] Elneweihi, AF; Schweitzer, EO; III, Feltis, MW. "Negative-sequence overcurrent element application and coordination in distribution protection," IEEE Power engineering society, *PES Summer Meeting, Seattle*, WA, July 12-16, 1992.

[15] IEEE Std C37.112-1996, IEEE *Standard Inverse-Time Characteristic Equations for Overcurrent Relays*, September 1996.

[16] IEEE Std. C37.113-1999, *Guide for protective relay applications to transmission lines,* September 1999.

[17] Sachdev, MS; Singh, J; Fleming, RJ. "Mathematical models representing time-current characteristics of overcurrent relays for computer application," *IEEE Paper* A78 131-5, January 1978.

[18] IEC Standard 255-4, *Single Input Energizing Measuring Relays with Dependent Specified Time*, IEC Publication 255-4, First Edition, 1976.

In: Electric Power Systems in Transition   ISBN: 978-1-61668-985-8
Editors: Olivia E. Robinson, pp. 197-217   © 2010 Nova Science Publishers, Inc.

Chapter 5

POWER SYSTEMS STATE ESTIMATION

Jizhong Zhu[*]

Chongqing University, Shapingba, Chongqing 400044, China

ABSTRACT

The electric power industry is undergoing massive changes around the world. Despite the changes with different structures, market rules, and uncertainties, an energy management system (EMS) control center must always be in place to maintain the security, reliability, and quality of electric service [1-2]. It means that EMS in the open energy market must respond quickly, reliably and efficiently to the market changes.

In this case, the function of state estimation (SE) is becoming more important, because it is the primary tool for monitoring and control based on the real-time data received from the measurement units. State estimation takes all the telemetry seen so far and uses it to determine the underlying behavior of the system at any point in time. It contributes to excluding errors in meters and compensating meter deficiency due to some faults with other measurements. As a result, it is possible for system operators to understand the state of the system through state estimation appropriately. This chapter introduces the methods of state estimation including SE with phasor measurement unit (PMU) and harmonic state estimation.

1. INTRODUCTION

Security operation of power system requires reliable real-time estimates of system state. Unfortunately, the complete state is not always observable. State estimation takes all the telemetry seen so far and uses it to determine the underlying behavior of the system at any point in time. It contributes to excluding errors in meters and compensating meter deficiency due to some faults with other measurements. As a result, it is possible for system operators to understand the state of the system through state estimation appropriately [3].

---

[*] Corresponding author: Email: jizhong.zhu@ieee.org

The state estimations typically include the following functions [4]:

- Topology processor: Gathers status data about the circuit breakers and switches, and configures the one-line diagram of the system.
- Observability analysis: Determines if a state estimation solution for the entire system can be obtained using the available set of measurements. Identifies the unobservable branches and the observable islands in the system if any exist.
- State estimation solution: Determines the optimal estimate for the system state, which is composed of complex bus voltages in the entire power system, based on the network model and the gathered measurements from the system. Also provides the best estimates for all the line flows, loads, transformer taps, and generator outputs.
- Bad data processing: Detects the existence of gross errors in the measurement set. Identifies and eliminates bad measurements provided that there is enough redundancy in the measurement configuration.
- Parameter and structural error processing: Estimates various network parameters, such as transmission line model parameters, tap changing transformer parameters, shunt capacitor or reactor parameters. Detects structural error in the network configuration and identifies the erroneous breaker status provided that there is enough measurement redundancy.

Among these functions/categories, the most important functions in state estimation are: bad data processing and topological observability analysis [5-19]. The bad data processing involves the identification of bad measurements and determination of critical measurements. The ability of a state estimator to detect and identify bad measurements depends both on the robustness of the state estimation algorithm and on the configuration of the measurement system such as number, type, and location of the measurements. It is well known that if a single bad data exists, the measurement with the largest normalized residual is the bad measurement. But this may be questionable if the erroneous measurement is an element of a minimally dependent set of measurements since all these measurements have absolutely equal normalized residuals. These minimally dependent set of measurements are called bad data groups.

Observability analysis in state estimation is related to system configurations or network topology. It focuses on whether state estimation can be carried out with a given set of measurements. Since power system topologies often vary along the daily system scheduling, there is a possibility that measurements are lost due to some faults. Thus, it is necessary to analyze the relationship between meter location and system configurations each time they are changed, or some measurements are lost or modified. Especially, the critical measurements play key role in observability analysis. The objectives of the topological observability consist of the following:

(1) to judge if a network is observable or not;
(2) to identify the observable area when a network is not observable;
(3) to identify the branches where pseudo-measurements are necessary to make the network observable when the network is not observable.

Recently, synchronized phasor measurement techniques are introduced in the field of power systems. The phasor measurement unit (PMU) is a power system device capable of measuring the synchronized voltage and current phasor in a power system. Synchronicity among phasor measurement units (PMUs) is achieved by same-time sampling of voltage and current waveforms using a common synchronizing signal from the global positioning satellite (GPS) [21]. In order to obtain simultaneous measurements of different buses it is necessary to synchronize sampling clocks at different locations. A PMU that can increase the confidence in the state estimation result is practically introduced via the use of synchronized measurements. Wide-area information from properly distributed PMUs enables the effective assessment of the dynamic performance of the power system and multi-area state estimation with independently operating system areas [22-29].

2. STATE ESTIMATION MODEL

Power system state estimation derives a real-time model through the received data from a redundant measurement set. Different kinds of methods about state estimation are introduced in [4]. Among them, the weighted least squares (WLS) state estimation methods are widely used. WLS state estimation minimizes the weighted sum of squares of the residuals.

Consider the set of measurements given by the vector z

$$z = \begin{bmatrix} z_1 \\ z_2 \\ \vdots \\ z_m \end{bmatrix} = \begin{bmatrix} h_1(x_1, x_2, ..., x_n) \\ h_2(x_1, x_2, ..., x_n) \\ \vdots \\ h_m(x_1, x_2, ..., x_n) \end{bmatrix} + \begin{bmatrix} e_1 \\ e_2 \\ \vdots \\ e_m \end{bmatrix} = h(x) = e \tag{1}$$

Where,
x: the system state vector;
e: the vector of measurement errors;
h: the nonlinear function relating measurement i to the state vector x.

There are three most commonly used measurement types in power system state estimation. They are the bus power injections, the line power flows and the bus voltage magnitudes. These measurement equations can be expressed using the state variables, which are given below from power flow equations:
1. Real and reactive power injection at bus i:

$$P_i = V_i \sum_{j=1}^{n} V_j (G_{ij} \cos \theta_{ij} + B_{ij} \sin \theta_{ij}) \tag{2}$$

$$Q_i = V_i \sum_{j=1}^{n} V_j (G_{ij} \sin \theta_{ij} - B_{ij} \cos \theta_{ij}) \tag{3}$$

2. Real and reactive power flow from bus i to bus j:

$$P_{ij} = V_i^2 (G_{si} + G_{ij}) - V_i V_j (G_{ij} \cos\theta_{ij} + B_{ij} \sin\theta_{ij})$$ (4)

$$Q_{ij} = -V_i^2 (B_{si} + B_{ij}) - V_i V_j (G_{ij} \sin\theta_{ij} - B_{ij} \cos\theta_{ij})$$ (5)

Where,

V_i : the voltage magnitude at bus i;

θ_i : the voltage angle at bus i;

P_i : the real power injection at bus i;

Q_i : the reactive power injection at bus i;

θ_{ij} : the voltage angle different between bus i and j;

P_{ij} : the real power flow from bus i to bus j;

Q_{ij} : the reactive power flow from bus i to bus j;

G_{ij} : the conductance of branch ij;

B_{ij} : the susceptance of branch ij.

Consider a system having N buses, the state vector will have $(2N - 1)$ elements, N bus voltage magnitudes and $(N - 1)$ phase angles. The state vector x will have the following form assuming bus 1 is selected as the reference:

$$x^T = [\theta_2 \theta_3 ... \theta_N V_1 V_2 ... V_N]$$

Let $E(e)$ denote the expected value of e, with the following assumptions:

$$E(e_i) = 0, \quad i = 1,...,m$$ (6)

$$E(e_i e_j) = 0$$ (7)

The measurement errors are assumed to be independent and their covariance matrix is given by a diagonal matrix R :

$$Cov(e) = E[e \cdot e^T] = R = diag\{\sigma_1^2, \sigma_2^2, ..., \sigma_m^2\}$$ (8)

The standard deviation σ_i of each measurement i is computed to reflect the expected accuracy of the corresponding meter used.

The WLS estimator will minimize the following objective function:

$$J(x) = \sum_{i=1}^{m} (z_i - h_i(x))^2 / R_{ii}$$
$$= [z - h(x)]^T R^{-1} [z - h(x)] \tag{9}$$

At the minimum value of the objective function, the first-order optimality conditions have to be satisfied. These can be expressed in compact form as follows.

$$g(x) = \frac{\partial J(x)}{\partial x} = -H^T(x)R^{-1}[z - h(x)] = 0 \tag{10}$$

where

$$H(x) = \left[\frac{\partial h(x)}{\partial x}\right] = \begin{bmatrix} \dfrac{\partial P_i}{\partial \theta} & \dfrac{\partial P_i}{\partial V} \\[2mm] \dfrac{\partial P_{ij}}{\partial \theta} & \dfrac{\partial P_{ij}}{\partial V} \\[2mm] \dfrac{\partial Q_i}{\partial \theta} & \dfrac{\partial Q_i}{\partial V} \\[2mm] \dfrac{\partial Q_{ij}}{\partial \theta} & \dfrac{\partial Q_{ij}}{\partial V} \\[2mm] \dfrac{\partial I_{ij}}{\partial \theta} & \dfrac{\partial I_{ij}}{\partial V} \\[2mm] 0 & \dfrac{\partial V_{ij}}{\partial V} \end{bmatrix} \tag{11}$$

is the measurement Jacobian matrix. The expressions of each partition can be computed using equations (2) – (5).

The non-linear function *g(x)* can be expanded into its Taylor series around the state vector x^k, that is

$$g(x) = g(x^k) + G(x^k)(x - x^k) + \ldots = 0 \tag{12}$$

Neglecting the higher order terms in the above expression, an iterative solution scheme known as the Gauss-Newton method is used to solve the following equation:

$$x^{k+1} = x^k - [G(x^k)]^{-1} \cdot g(x^k) \tag{13}$$

where,

k: the iteration index;

x^k: the solution vector at iteration k:

$G(x)$: the gain matrix, which is expressed as follows.

$$G(x^k) = \frac{\partial g(x^k)}{\partial x} = H^T(x^k)R^{-1}H(x^k)$$

(14)

$$g(x^k) = -H^T(x^k)R^{-1}\big(z - h(x^k)\big)$$

(15)

Generally, the gain matrix $G(x)$ is sparse, positive definite and symmetric provided that the system is fully observable. It can be decomposed into its triangular factors. At each iteration k, the following sparse linear set of equations are solved using

$$\big[G(x^k)\big]\Delta x^{k+1} = H^T(x^k)R^{-1}\big[z - h(x^k)\big]$$
$$\Delta x^k = x^{k+1} - x^k$$

(16)

3. STATE ESTIMATION ALGORITHMS

3.1. WLS Algorithm for State Estimation

Equation (16) is called as Normal Equation. WLS state estimation uses the iterative solution of the Normal Equation. Iterations start at an initial guess x^0 which is typically chosen as the flat start, i.e. all bus voltages are assumed to be 1.0 per unit and in phase with each other. The iterative solution algorithm for WLS state estimation can be summarized as below.

1)   Initially set the iteration counter $k = 0$, and set the maximum iteration number k_{max}.
2)   If $k > k_{max}$, then terminate the iterations.
3)   Calculate the measurement function $h(x^k)$, the measurement Jacobian $H(x^k)$ and the gain matrix $G(x^k)$.
4)   Solve equation (13) to get Δx^k.
5)   Check for convergence, i.e., $\max|\Delta x^k| \le \varepsilon$. If yes, stop. Otherwise, go to next.
6)   Update $x^{k+1} = x^k + \Delta x^k$, $k \Leftarrow k+1$, and go to step 2.

3.2. Topological Observability Analysis [3]

3.2.1 Formulation of topological observability [8-16]

The minimum spanning tree (MST) technique in graph theory is used for topological observability analysis. For an n node system with m measurements, the linearized observation equation can be written as

$$Z = HX + \beta \qquad (17)$$

Where
Z: measurement vector with dimension $m \times 1$;
H: measurement matrix with dimension $m \times (2n - 1)$;
X: voltage vector with dimension $(2n-1) \times 1$;
β: noise vector with dimension $m \times 1$;
m: number of measurements;
n: number of nodes.

To solve the equation (17), the number of measurements should be greater than the number of nodes ($m > n$). It is necessary that the measurement matrix is independent with respect to the n columns. Since the state vector X has $(2n - 1)$ dimension, the network is said to be observable or algebraically observable if the measurement matrix meets

$$Rank(H) = 2n - 1 \qquad (18)$$

Where, rank($\cdot$) stands for the rank of matrix.

If we express voltage vector in equation (17) in polar coordinates and apply $P - \theta / Q - V$ decoupling principle to the equation, we get the following two equations

$$Z_P = H_{P\theta} X_\theta + \beta_P \qquad (19a)$$

$$Z_Q = H_{QV} X_V + \beta_Q \qquad (19b)$$

Where
Z_P: the real power measurement vector ($m_P \times 1$);
Z_Q: the reactive power/voltage measurement vector ($m_Q \times 1$);
X_θ: the voltage phase angle vector (($n-1) \times 1$);
X_V: the voltage magnitude vector ($n \times 1$);
$H_{P\theta}$:   the measurement matrix with respect to X_θ ($m_P \times (n-1)$);
H_{QV}:   the measurement matrix with respect to X_V ($m_Q \times n$);
β_P: the real power noise vector ($m_P \times 1$);
β_Q: the reactive power/voltage noise vector ($m_Q \times 1$);
m_P: the number of real power measurements;
m_Q: the number of reactive power/voltage measurements.

The network is said to be $P - \theta$ observable or $P - \theta$ algebraically observable if the following equation holds:

$$Rank(H_{P\theta}) = n - 1 \qquad (20)$$

In the same way, the network is said to be $Q - V$ observable or $Q - V$ algebraically observable if the following equation holds:

$$Rank(H_{QV}) = n \qquad (21)$$

According to graph theory, we can use a weighted graph to represent meter placement in the be $P - \theta$ network whose branch weights are assigned by whether MW meters are located on the branches or not. Let $G_{P\theta}$ be the weighted graph, which means a matrix representing MW meter allocation rather than the measurement matrix $H_{P\theta}$ itself. The network is said to be $P - \theta$ topologically observable if the following equation holds:

$$Rank(G_{P\theta}) = n - 1 \qquad (22)$$

Similarly for a weighted graph G_{QV}, representing MVAr meter allocation, the network is said to be $Q - V$ topologically observable if the following equation holds:

$$Rank(G_{QV}) = n \qquad (23)$$

Assume that the real and reactive power measurements are taken in pairs and the voltage magnitude at the reference node is measured in the reference node. Consequently, solving the $P - \theta$ topologically observable is equivalent to solving $Q - V$ topologically observable. It has been proved that the conditions are that a network is topologically observable if there is a spanning tree of full rank on graph $G_{P\theta}$ [16-19].

Before describing the formulation for general measurements, we discuss a simple case to understand the outline of the minimum spanning tree (MST) based approach. This is based on the following assumptions:

(1) As mentioned before, the real and reactive power measurements are in pairs and the voltage magnitude at the reference node is measured in the reference node.
(2) The real and reactive power measurements consist of line flow only.

As a result, examining topologically observable in the $P - \theta$ network is equivalent to doing it in the $Q - V$ network. For simplicity, the nodal power measurements and more than two voltage measurements are not considered.

For the above mentioned weighted graph $G_{P\theta}$ representing network meter placement, the branch weights of $G_{P\theta}$ are determined by whether meters are located at the branches or the transmission lines in the power system.

$$C_{bi} = \begin{cases} 0, & \text{if branch } i \text{ has meters} \\ 1, & \text{if branch } i \text{ has no meter} \end{cases} \qquad (24)$$

Let W_{bi} be the status of the branch in the operation of power system, that is

$$W_{bi} = \begin{cases} 0, \textit{if branch i is in outage} \\ 1, \textit{if branch i is in service} \end{cases} \qquad (25)$$

We mentioned that the network is topologically observable if the spanning tree is of full rank. Thus, the problem of topological observability is translated into evaluating the minimum spanning tree $T_{P\theta}$ in graph $G_{P\theta}$ with available weights $(W_{bi}C_{bi})$. According to graph theory, the minimum spanning tree (MST) means the spanning tree whose sum of the weights is minimum among the existing spanning trees on graph $G_{P\theta}$. Let $F(T_{P\theta})$ be the sum of the available branch weights in the spanning tree. The problem of topological observability can be solved as follows.

$$\min F(T_{P\theta}) = \sum_{bi \in T_{P\theta}} W_{bi} C_{bi} \qquad (26)$$

According to the above equation, the network topological observability is judged as below.

1) Observable network

   If all branches are in service, the minimum spanning tree (MST) obtained from equation (26) consists of the $(n-1)$ branches with $C_{bi} = 0$. Since the network is topologically observable, the line flow measurements correspond to the node angles/voltages in a one-to-one way. Therefore, the sum $F(T_{P\theta})$ is represented as follows.

   $$F(T_{P\theta}) = 0 \qquad (27)$$

2) Unobservable network

   Assume all branches are in service. If the minimum spanning tree (MST) has at least one branch with $C_{bi} = 1$, the network is unobservable. In this case, the sum $F(T_{P\theta})$ is given as follows.

   $$F(T_{P\theta}) > 0 \qquad (28)$$

In the state estimation of power network, firstly it needs to judge if a network is topologically observable. If the network (especially a large scale power system) is topologically observable, the following measures will be adopted:

1) to identify the maximum observable island or area.
2) To identify the branches where pseudo-measurements should be placed to recover topological observability.

In this way, the network is decomposed into several observable sub-networks which are called the observable islands (or areas). The maximum observable island (or area) is the largest sub-network among them. That is the maximum observable island (or area) is the sub-network that consists of the most branches with good measurements among the obtained MST. It is efficient to make use of the maximum observable island (or area) in order to make whole network topologically observable through placing the pseudo-measurements on some branches without measurements in unobservable sub-networks.

3.2.2 Augmented graph for observability analysis

In the practice, in addition to line flow measurements, it is necessary to consider nodal power injection and more than one voltage measurements, which were neglected in section 3.2.1. This section introduces the method of augmented graph to deal with nodal power injection and more than one voltage measurements. The approach is based on the fact that voltage measurements are changed into the equivalent reactive power flow as well as the nodal injection measurements are represented as the equivalent power flows from the ground to the existing nodes. Consequently, the augmented graphs are created by adding one new node on graphs $G_{P\theta}$ and G_{QV} to connecting all the existing nodes. The $P - \theta$ and $Q - V$ topological observability are evaluated by finding the MST in the augmented graphs, respectively.

Let n be the number of nodes in the network. Since the voltage measurements are transformed into the equivalent line flow measurements that connect the ground, which is a new node $(n+1)$, the $Q - V$ topological observability requires n branches with measurements corresponding to each node. The sum of branch weights in MST becomes n in the observable system since the sum of nodes is equal to $(n+1)$.

Similarly, the nodal reactive power injection measurement at node k can be handled as reactive line flow measurement from the ground to node k. The same procedure basically can be utilized in the MST technique for $Q - V$ topological observability analysis. It is noted that the voltage measurement covers its own node while the nodal injection measurement at node k may be assigned to the neighboring nodes in an augmented graph if a neighboring node requires a measurement to keep the observability. That is the nodal injection measurement has conditions that the equivalent flow measurement is allowed to one of the neighboring nodes, which makes the injection measurements are flexible to some extent. According the above analysis, the $Q - V$ topological observability can be evaluated by the following steps [10]:

Step 1: Transform voltage and nodal injection measurements into the equivalent line flow measurement. Form the augmented graph.
Step 2: Evaluate the minimum spanning tree in the augmented graph.
Step 3: If the MST is found and all branches have measurements, then stop. The $Q - V$ network is topologically observable. Else go to step 4.
Step 4: Find the isolated nodes and the neighboring nodes.
Step 5: If other nodal measurement can be assigned to the isolated nodes as line flow measurement, then update the MST and go back to step 3. Otherwise, stop. The $Q - V$ network is not topologically observable.

3.3. Identification of Bad Measurement Data [3]

From the analysis in previous sections if minimum spanning tree (MST) of network contains the measurement in each branch in MST, the network is topologically observable. However, we didn't check the quality of the measurements. The network may be unobservable if there exists a bad measurement in MST. This section discusses the identification of bad measurement data. The ability of a state estimator to detect and identify bad measurements depends both on the robustness of the state estimation algorithm and on the configuration of the measurement system (number, type and location of the measurements). It is well know that if a single bad data exists, the measurement with the largest normalized residual is the bad measurement. This however is not true if the erroneous measurement is an element of a minimally dependent set of measurements since all these measurements have absolutely equal normalized residuals. This is the reason why minimally dependent sets of measurements are also called as bad data groups.

3.3.1. Properties and classification of bad data groups

This section presents the definitions and theorems related to bad data groups. These theorems apply to both $P - \theta$ and $Q - V$ problems by considering voltage measurements as equivalent reactive flow measurements of fictitious branches connecting the existing nodes and the ground [17-19].

Definition 1: A measurement is defined as critical if its suppression from the measurement set makes the system unobservable.

Error on a critical measurement can be neither detected nor identified.

Definition 2: A set of noncritical measurements is defined as a minimally dependent set of measurements (or bad data groups of measurements) if the removal of any measurement from this set makes the remaining measurements of the set critical.

Definition 3: Error on a noncritical measurement spreads only to the residuals of the measurements wholly contained in an error residual spread area.

Definition 4: A flow island is defined as a set of nodes with the property that for any two nodes of this set a path of flow measured branches always exists between these two nodes.

In definition 4, a branch is an equivalent representation of multiple lines between two nodes (or buses). A branch is defined as "flow measured branch" if at least one of the individual lines is flow measured. According to definition 4, we have the following definition.

Definition 5: Injections at internal or boundary nodes of a flow island will be refereed as internal or boundary injection measurements respectively.

Theorem 1: The measurements of a bad data group have all absolutely equal normalized residuals.

Theorem 2: The measurements of a bad data group are all wholly contained into one error residual spread area.

Theorem 3: Two or more internal injection measurements cannot belong to the same bad data group of measurements.

Theorem 4: An internal and one or more boundary injection measurements cannot belong to the same bad data group of measurements.

Any combination of injection and/or flow measurements is a candidate bad data group of measurements. By excluding the combinations mentioned in Theorems 3 and 4, only the following four classes are defined as bad data groups of measurements:

Class 1: Bad data group of two flow measurements belonging to the same flow measured branch.

Class 2: Bad data group of measurements consisting of one internal injection measurement and one or more flow measurements.

Class 3: Bad data group of one or more boundary injection measurements and possibly of flow measurements.

Class 4: Bad data group of flow measurements belonging to different flow measured branches.

3.3.2. Criticality of flow measurements based on graph theory

Criticality of flow measurements is evaluated based on graph theory. Let G be a connected graph with n nodes and m branches. A connected sub-graph T is a tree of G. There are $n-1$ branches in the tree T. The $m-n+1$ branches of G not in T are called links or co-tree branches. The reduced node-to-branch incidence matrix of G is defined as A with $(n-1)\times m$, in which the element is defined as follows.

$$a_{ij} = \begin{cases} 1, & \textit{if node i is sending end of branch j} \\ -1, & \textit{if node i is receiving end of branch j} \\ 0, & \textit{if branch j is not connected to node i} \end{cases}$$
(29)

Each link branch together with the unique path of tree branches between its nodes forms a fundamental loop for that link branch. The direction of the link tree branch defines the loop direction. The element of the fundamental loop matrix B with $(n-m+1)\times m$ is defined as

$$b_{ij} = \begin{cases} 1, & \textit{if branch j is in loop i and their directions agree} \\ -1, & \textit{if branch j is in loop i and their directions oppose} \\ 0, & \textit{if branch j is not in loop i} \end{cases}$$
(30)

Each tree branch of T together with some (possibly none) link branches forms a fundamental cut-set for that tree branch. The direction of the tree branch defines the direction of cut-set. We use D represent a cut-set matrix with $(n–1)\times m$, whose element is defined as

$$d_{ij} = \begin{cases} 1, & \textit{if branch j belongs to cutset i and they have the same direction} \\ -1, & \textit{if branch j belongs to cutset i and they have the opposite direction} \\ 0, & \textit{if branch j does not belong to cutset i} \end{cases} \quad (31)$$

Assume the matrices A, B and D have the maximal number of linearly independent rows and can be partitioned as:

$$A = \begin{bmatrix} A_T & A_L \end{bmatrix} \quad (32)$$

$$B = \begin{bmatrix} B_T & I_L \end{bmatrix} \quad (33)$$

$$D = \begin{bmatrix} I_T & D_L \end{bmatrix} \quad (34)$$

Where the columns of A_T, B_T and I_T correspond to tree branches, and the columns of A_L, I_L and D_L correspond to link branches. According to graph theory, we have

$$\det(A_T) \neq 0 \quad (35)$$

$$B_T = -D_L^T \quad (36)$$

$$D = \begin{bmatrix} I_T & D_L \end{bmatrix} = A_T^{-1} \begin{bmatrix} A_T & A_L \end{bmatrix} = A_T^{-1} A \quad (37)$$

It means that if we perform elementary row operations on A to reduce A_T to an identity matrix, the resulting matrix is D.

Let G be the graph consisting of flow measured branches and branches incident to non-critical injection measurements. Let D_L be the sub-matrix of the fundamental cut-set matrix D that corresponds to link branches. For the Jacobian matrix H_S related to one connected sub-graph G_S of G it holds:

$$H_S = MA^T \quad (38)$$

Where,
A: the reduced node-to-branch incidence matrix for the sub-graph G_S.
M: the measurement-to-branch incidence matrix, whose element is defined as below.

$$m_{ij} = \begin{cases} 1, & \text{if the } i-\text{th measurement is incident to the sending node of the } j-\text{th branch} \\ -1, & \text{if the } i-\text{th measurement is incident to the receiving node of the } j-\text{th branch} \\ 0, & \text{elsewhere} \end{cases}$$

For each connected sub-graph G_S of G, we select a tree consisting of flow measured branches and unmeasured branches connecting flow islands. Based on equations (36) and (37), we get

$$H_S A_T^{-T} = MD^T = M \begin{bmatrix} I_T \\ D_L^T \end{bmatrix} = M \begin{bmatrix} I_T \\ -B_T \end{bmatrix} \tag{39}$$

Rows of H_S corresponding to critical flow measurements are linearly independent of the other rows of H_S. Since A_T^{-T} is nonsingular, linear dependencies among rows of H_S are equivalent to those of $H_S A_T^{-T}$ or equivalent to those of $M \begin{bmatrix} I_T \\ -B_T \end{bmatrix}$. Therefore, we get:

(a) A flow measurement incident to the i-th tree branch corresponds to a row of $H_S A_T^{-T}$, which is in the following form:

$$\pm e_i^T \tag{40}$$

Where e_i is a vector of dimension equal to the number of tree branches, having all its elements equal to zero except the i-th element that is equal to 1.

(b) A flow measurement incident to the j-th link branch corresponds to a row of $H_S A_T^{-T}$, which is in the following form:

$$\sum_{i \in l(j)} a_i e_i^T \tag{41}$$

Where $a_i = \pm 1$ and $l(j)$ is the unique set of flow measured tree branches forming loop with the j-th link branch.

(c) An injection measurement at node k corresponds to a row of $H_S A_T^{-T}$, which is in the following form:

$$\sum_{i \in t(k)} a_i e_i^T + \sum_{j \in c(k)} b_j \sum_{i \in l(j)} a_i e_i^T \tag{42}$$

Where $a_i = \pm 1$, $b_j = \pm 1$ and $t(k)$ and $c(k)$ are sets of tree and link branches respectively incident to node k.

From equations (40) – (42), we have the following criteria to examine if a flow measurement is critical.

A flow measurement is critical if, for the related flow measured branch, the following three criticality conditions are satisfied:

(1) The flow measured branch is a tree branch containing only one flow measurement.
(2) The flow measured branch is not incident to any non-critical injection measurement.
(3) The row of D_L corresponding to the flow measured branch has all its elements equal to zero.

If one measurement is eliminated from the measurement set, the resulting critical flow measurements are determined by the following procedure:

The resulting critical flow measurements, after the elimination of a non-critical measurement, are determined by testing the criticality conditions, without examining the columns of D_L corresponding to the link branches incident only to the eliminated measurement or to the resulting critical injection measurements.

According the above mentioned theory, the identification of bad data group of measurements can be conducted using the following steps.

Step 1: Form a graph.
Step 2: Identify flow islands.
Step 3: Compute the matrix of the reduced network and determine critical injection measurements.
Step 4: Create a sub-graph consisting of flow measured branches and branches incident to non-critical injection measurements.
Step 5: Compute matrix and determine critical flow measurements.
Step 6: Define the error residua spread areas based on branches incident to non-critical measurements.
Step 7: For each error residual spread area, determine the bad data groups of measurements.

4. HARMONIC STATE ESTIMATION

The harmonic state estimation (HSE) in power system estimates the harmonic state in whole network according to the measured values of limited points [30-31]. The problem of harmonic state estimation was first proposed by Heydt in 1989 [30], where HSE is handled as an inverse problem of harmonic power flow, and the adopted algorithm for estimating the harmonic state and identifying the harmonic source is least square estimation. The HSE mathematical model can be expressed as:

$$Hx = \left[H_{\text{true}} + v \right] x = z_{\text{true}} + \varepsilon = z \tag{43}$$

Where, H is the parameter matrix ($m \times n$) containing the relationship between the measured values and the state variables. x is the undetermined state vector ($n \times 1$). H_{true} is the parameter matrix containing the truth values. v is the parameter error matrix ($m \times n$). z_{true} is the measurement vector ($m \times 1$) related to the truth values, which are unknown variables. ε is the measurement error vector ($m \times 1$). z is the measurement vector ($m \times 1$).

The harmonic state estimation equation (43) is an overdetermined linear system, which can be solved by total least square (TLS) method. The estimation method of TLS is trying to estimate the noise matrix v and the noise vector ε to meet the exact solution of linear system. Select v and ε to make a minimum of

$$\left\| \begin{bmatrix} \varepsilon & v \end{bmatrix} \right\|_F^2 = \sum_{i=1}^{m} \varepsilon_i^2 + \sum_{i=1}^{m} \sum_{j=1}^{n} v_{ij}^2 \tag{44}$$

where, v_{ij} is the element of matrix v, and ε_i is the i-th element of ε.

Let $\bar{H} = [z, H]$，equation (43) can be written as follows:

$$(\bar{H} - \begin{bmatrix} \varepsilon & v \end{bmatrix}) \begin{bmatrix} -1 \\ x_{\text{TLS}} \end{bmatrix} = 0 \tag{45}$$

where, 0 is an m-dimension vector in which all the elements are 0.

In general, because of the existence of noise, the augmented matrix $\bar{H}$ is full rank. If $m > n+1$, the rank of matrix $\bar{H}$ is ($n+1$). Using the singular value decomposition (SVD) method, the matrix can be expanded as follows:

$$\bar{H} = [z, H] = \sum_{i=1}^{n+1} \sigma_i u_i w_i^T \tag{46}$$

where, σ_i is the singular value of matrix $\bar{H}$, arranging in the decreasing order of values, u_i and w_i are left singular vector containing m elements and right singular vector containing ($n+1$) elements, respectively.

Let's define rank approximate value as the minimum of the sum of various perturbations in matrix $\bar{H}$. It is given as follows [4]:

$$\hat{H} = \sum_{i=1}^{n} \sigma_i u_i w_i^T \tag{47}$$

Moreover, error matrix $\begin{bmatrix} \varepsilon & v \end{bmatrix}$ is given as follows :

$$[\boldsymbol{\varepsilon} \quad \boldsymbol{v}] = \sigma_{n+1} \boldsymbol{u}_{n+1} \boldsymbol{w}_{n+1}^{T} \tag{48}$$

From equations (45) and (47), we can get

$$\begin{bmatrix} -1 \\ \boldsymbol{x}_{\mathrm{TLS}} \end{bmatrix} = -\frac{\boldsymbol{w}_{n+1}}{w_{n+1,1}} \tag{49}$$

When the noise sequence meets the central limit theorem conditions of independent identical distribution, the standard TLS estimation is unbiased.

For the h-th harmonic, the harmonic power flow equation can be expressed as follows:

$$Y(h)\dot{V}(h) = \dot{I}(h) \tag{50}$$

In this section, the node harmonic voltage vector $\dot{V}(h)$ is used as the measuring point, and the node harmonic injection current $\dot{I}(h)$ is used as the state variables. If different measurements and state variables are selected, the similar state estimation expression can still be obtained through the matrix transformation. If the parameters error and measurement error are considered simultaneously, that is, the harmonic impedance matrix error and harmonic node voltage measurement error are considered together, the relationship between the measurements and the state variables can be obtained as follows:

$$\dot{V}_{\mathrm{true}}(h) + \boldsymbol{\varepsilon}(h) = (\boldsymbol{Z}_{\mathrm{true}}(h) + \boldsymbol{v}(h))\dot{I}(h) \tag{51}$$

According to equations (45) – (49), the estimated value of the node harmonic injection current, $\dot{I}_{\mathrm{TLS}}(h)$, can be computed. Thus, the harmonic state estimation is solved.

5. PMU BASED STATE ESTIMATION

One PMU can measure not only the voltage phasor, but also the current phasors. Generally, those measurements received from PMUs are more accurate with small variances compared to the variances of the conventional measurements. Therefore, inclusion of PMU measurements is expected to produce more accurate estimates.

If the measurement set is composed of only voltages and currents measured by PMUs, the state estimation can be formulated as a linear problem.

$$Z = \begin{bmatrix} V_B \\ I_L \end{bmatrix} + \begin{bmatrix} \varepsilon_B \\ \varepsilon_L \end{bmatrix} \tag{52}$$

Where,

V_B :the true bus voltage vector;

I_L :the true line current vector;

ε_B :the error vector of bus voltage estimation;

ε_L :the error vector of line current estimation.

The true line current vector I_L can be expressed by the true bus voltage vector V_B, Then we get

$$Z = \begin{bmatrix} I \\ YA^T + Y_S \end{bmatrix} V_B + \begin{bmatrix} \varepsilon_B \\ \varepsilon_L \end{bmatrix} = HV_B + \varepsilon \tag{53}$$

The measurement errors were assumed to be un-correlated with zero mean defined by the diagonal covariance D. The WLS method can be used to solve (54) for V_B.

Other state estimation modes were based on integrating the PMU angle measurement, that is

$$Z = \begin{bmatrix} Z_A \\ Z_R \end{bmatrix} \tag{54}$$

where

$$Z_A = \begin{bmatrix} P_{ij} \\ P_i \\ \theta_i \end{bmatrix}; \quad Z_R = \begin{bmatrix} Q_{ij} \\ Q_i \\ V_i \end{bmatrix} \tag{55}$$

And θ_i is the direct angle measurement with PMU.

It is noted that the reference of PMU measurements is the one associated with the timing signal from the GPS, say θ_{ref} while the state from the traditional SE are referred to the angle of one of the buses, say δ_{slack}. It is necessary to align these two references. Otherwise the results would not make sense. If we align the WLS and the PMU references, we should measure the same angle of the voltage phasor at bus i. To accomplish this, we must adjust the PMU angle reference by a certain amount Φ to align it with the SE reference. For the bus i, the angle relationship between PMU and traditional SE measurement can be expressed as follows [21].

$$\theta_i + \theta_{ref} = \delta_i + \delta_{slack} \tag{56}$$

This means that the angle required to align the references is

$$\Phi = \theta_{ref} - \delta_{slack} = \delta_i - \theta_i \tag{57}$$

If we have m sets of PMU measurements then we can take the average so that a good estimate of Φ (the phase shift between PMU reference and the slack bus angle) is

$$\Phi = \sum_{i=1}^{m} \frac{\delta_i - \theta_i}{m} \tag{58}$$

REFERENCES

[1]   Dy-Liyacco, TE. "Control Centers Are Here to Stay," *IEEE Computer Applications in Power*, Vol.15, No.4, 18-23, 2002.

[2]   Zhu, JZ. *Optimization of Power System Operation*, New York: Wiley-IEEE Press, August 2009.

[3]   Zhu, JZ. *Power System Applications of Graph Theory*, New York: Nova Science Publishers, September 2009.

[4]   Abur, A; Expósito, AG. *Power System State Estimation Theory and Implementation*, New York: Wiley-IEEE Press, August 2004.

[5]   Schweppe, FC; Wildes, J. "Power system static-state estimation, part I: exact model," *IEEE Trans. on PAS,* Vol.PAS-89, 120-125, January 1970.

[6]   Schweppe, FC; Rom, DB. "Power system static-state estimation, part II: approximate model," *IEEE Trans. on PAS,* Vol.PAS-89, 125-130, January 1970

[7]   Schweppe, FC. "Power system static-state estimation, part III: implementation," *IEEE Trans. on PAS,* Vol.PAS-89, 130-135, January 1970.

[8]   Mori, H; Tsuzuki, S. "A fast method for topological observability analysis using a minimum spanning tree technique," *IEEE Trans. on Power System,* Vol.6, No.2, 491-500, May 1991.

[9]   Clements, KA; Wollenberg, BF. "An algorithm for observability determination in power system estimation," *IEEE PES Summer Meeting, Paper A*, 75 447-3, San Francisco, July 1975.

[10] Krumpholtz, GR. et al, "Power system observability: A practical algorithm using network topology," *IEEE Trans. on Power Systems*, Vol. 99, No.4, 1534-1542, 1980.

[11] Quintana, VH; Simoes-Costa, A; Mandel, A. "Power system observability using a direct graph-theoretic approach," *IEEE Trans. on Power Systems*, Vol. 101, No.3, 617-626, 1982.

[12] Monticelli, A; Wu, FF. "Network observability: Identification of observable islands and measurement placement," *IEEE Trans. on Power Systems*, Vol.104, No.5, 1035-1041, 1985.

[13] Horisberger, HP. "Observability analysis for power system with measurement deficiencies," Proc. *IFAC Symp. on Electric Energy Systems*, 51-58, Rio de Janeiro, Brazil, July 1985.

[14] Bargiela, A. et al., "Observability determination in power system state estimation," *IEEE Trans. on Power Systems*, Vol. PWRS-1, No.2, 108-114, 1986.

[15] Slutsker, IW; Scudder, JM. "Network observability analysis through measurement Jacobian matrix reduction," *IEEE Trans. on Power Systems*, Vol.PWRS-2, No.2 331-338, 1987.

[16] Clements, KA. et al., "State estimator measurement system reliability evaluation – an efficient algorithm based on topological Observability theory," *IEEE Trans. on Power Systems*, Vol.101, No.4, 997-1004, 1982

[17] Korres, GN; Contaxis, GC. "Identification and updating of minimally dependent sets of measurements in state estimation," *IEEE Trans. on Power System*, Vol.6, No.3, 999-1005, August 1991.

[18] Ayres, M; Haley, PH. "Bad data group in power system state estimation," *IEEE Trans. on Power Systems*, Vol. PWRS-1, No.1, 1-8, 1986.

[19] Contaxis, GC; Korres, GN. "A reduced model for power system observability analysis and restoration," *IEEE Trans. on Power System*, Vol.3, 1411-1417, 1988.

[20] Mili, L; Van Cutsem, T; Ribbens-Pavella, M. "Hypothesis testing identification: a new method for bad data analysis in power system state estimation," *IEEE Trans. on Power System*, Vol.103, 3239-3352, 1984.

[21] Nuqui, RF. "State estimation and voltage security monitoring using synchronized phasor measurements," *Dissertation*, 2001.

[22] Yoon, YJ. "Study of the utilization and benefits of phasor measurement units for large scale power system state estimation," *Thesis*, 2005.

[23] Thorp, JS; Phadke, AG; Karimi, KJ. "Real time voltage-phasor measurements for static state estimation", *IEEE Transactions on Power Apparatus and Systems*, vol. 104, 3098-3104, November 1985.

[24] Zivanovic, R; Cairns, C. "Implementation of PMU technology in state estimation: An overview", *IEEE AFRICON 4th*, University of Stellenbosch, South Africa, September 1996.

[25] Phadke, AG. *"Synchronized phasor measurements ~ a historical overview"*, in Proc. of the Transmission and Distribution Conference and Exhibition in Asia/Pacific, Yokohama, Japan, October 2002.

[26] Kamwa, I; Grondin, R. "PMU configuration for system dynamic performance measurement in large multiarea power systems", *IEEE Transactions on Power Systems*, vol. 17, 385-394, May 2002.

[27] Cho, KS; Shin, JR; Hyun, SH. " Optimal placement of phasor measurement units with GPS Receiver", in *IEEE Power Engineering Society Winter Meeting*, Columbus, Ohio, January 2001.

[28] Denegri, GB; Invernizzi, M; Milano, F; Fiorina, M; Scarpellini, P. *"A security oriented approach to PMU positioning for advanced monitoring of a transmission grid"*, in Proc. of the Int. Conference on Power System Technology, Kuming, China, October 2002.

[29] Milosevic, B; Begovic, M. "Nondominated sorting genetic algorithm for optimal phasor measurement placement", *IEEE Transactions on Power Systems*, vol. 18, 69-75, February 2003.

[30] Heydt, GT. "Identification of harmonic sources by a state estimation technique," *IEEE Trans. Power Delivery*, Vol. 4, No.1, 569-576, 1989.

[31] Meliopoulos, APS; Zhang, F; Zelingher, S. "Power system harmonic state estimation," *IEEE Trans. Power Delivery*, Vol.9, No.3, 1701-1709, 1994.

[32] Du, ZP; Arrillaga, J; Watson, N; Chen, S. "Identification of harmonic sources of power systems using state estimation," *IEE Proceedings - Generation, Transmission and Distribution*, Vol. 146, No.1, 7–12, 1999.

In: Electric Power Systems in Transition
Editors: Olivia E. Robinson, pp. 219-290

ISBN: 978-1-61668-985-8
© 2010 Nova Science Publishers, Inc.

Chapter 6

ADVANCED FAULT LOCATION TECHNIQUE FOR PARALLEL POWER TRANSMISSION LINES

Ning Kang and Yuan Liao

University of Kentucky, College of Engineering, Department of Electrical and Computer Engineering, Lexington, Kentucky

ABSTRACT

This article presents a general approach for locating any type of short circuit faults on a double-circuit transmission line. By making use of the bus impedance matrix technique, voltage measurements at one or two buses are utilized as inputs, which may be distant from the faulted section. The bus impedance matrix of each sequence network with addition of a fictitious bus at the fault point can be constructed as a function of fault location by drawing on network analysis. Fault location can then be obtained based on bus impedance matrix and boundary conditions of different fault types. It is assumed that the network data are available. Quite accurate results have been achieved based on simulation studies.

I. INTRODUCTION

Double-circuit transmission lines, or called parallel lines, have been adopted more and more in modern power systems because they can improve reliability and security of energy transmission. As is known, following the occurrence of a fault, it is important to promptly and accurately locate the fault and repair the faulted component to reduce outage time and loss of revenue.

A lot of research efforts have been made on double-circuit lines which resulted in various fault location algorithms. The authors of [1] propose an algorithm utilizing one-terminal voltage and current data, but the accuracy of fault location is influenced by fault resistance and the unsymmetrical arrangement of the transmission line. Eriksson et al. [2] employ phase voltages and currents from the near end of the faulted section as input signals and fully compensate the error introduced by the fault resistance. Based on the assumption that the line is homogeneous, [4] makes use of modal transformation [3] and the local terminal voltage and current to locate fault. Six voltage equations are constructed around the parallel loop for positive-, negative- and zero-sequence networks in [5], from which the fault location is solved. The input of [5] is the voltage and current phasors of one end of the faulted section.

A two terminal fault location method using unsynchronized voltage and current phasors based on a distributed parameter model is discussed by Johns et al [6]. Reference [7] relates the synchronized voltage and current phasors of the sending end and receiving end with ABCD parameters, from which the fault location can be derived. In [8], based on the differential component net decomposed from the original net, two voltage distributions along the line can be calculated from the unsynchronized two terminal currents by making use of the distributed parameter time-domain equivalent model and these two voltage distributions have the least difference at the fault point.

Generally speaking, existing algorithms require voltages and currents from one or two terminals of the faulted section or all the terminals of the network. For the scenario where only sparse measurements, which may be far away from the faulted section, are available, these methods are not suitable any more.

A fault location algorithm for single-circuit transmission line based on sparse measurements is presented in [9]. This article further develops novel one-bus and two-bus algorithms for double-circuit lines based on [9]. The measurements utilized for two-bus algorithm can be either synchronized or unsynchronized. No current measurements are needed. In this article, it is assumed that the network data are known and the network is transposed. The described fault location method is based on lumped parameter line model.

Section II reports the proposed fault location method, which is examined by simulation studies in Section III. Then Section IV summarizes the proposed method.

II. PROPOSED FAULT LOCATION METHOD

With addition of a fictitious bus at the fault point, the driving point impedance of the fault bus and the transfer impedances between this bus and other buses can be derived as functions of fault location. Based on the definition of bus impedance matrix, the voltage change at any bus during the fault can be formulated with respect to the corresponding transfer impedance. In conjunction with the boundary conditions of different fault types, the fault location can be obtained.

In the following sections, the bus impedance matrices with an additional fault bus for the positive-, negative- and zero-sequence network are constructed, respectively. Then the fault location algorithms are derived.

II.1. Construction of bus Impedance Matrix with Addition of the Fault Bus

The construction of bus impedance matrix with addition of the fault bus for the zero sequence network is first considered. The pre-fault zero sequence network of a sample power system with the faulted section extracted is shown in Figure 1, whose bus impedance matrix $Z_0^{(0)}$ is assumed to be already developed using well established techniques [10]. It should be noticed that throughout the article, 0, 1 or 2 in parenthesis as a superscript signifies zero-, positive- or negative-sequence quantities. Bus p and q are the two terminals of the faulted section. $z_a^{(0)}$, $z_b^{(0)}$ are respectively the total zero sequence self impedances of the two branches between bus p and q. The total mutual impedance between these two branches is $z_m^{(0)}$. The pre-fault network has n buses in total.

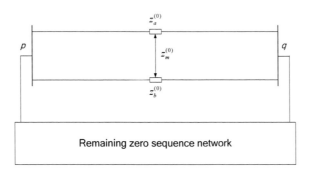

Figure 1. Pre-fault zero sequence network of a sample power system

Suppose the fault occurs on the branch with total impedance $z_b^{(0)}$. The fault point is denoted as r and let $r = n+1$. The network with an additional fault bus on the faulted line is modeled as in Figure 2, whose bus impedance matrix is $Z^{(0)}$. m is the per unit fault distance from bus p.

To formulate $Z^{(0)}$, suppose there is only one current source injected into a single bus l ($l = 1,2...n$), then the resulted voltages at bus k ($k = 1,2...n$) will be the same for the networks shown in Figure 1 and Figure 2. Based on the definition of bus impedance matrix, it is obtained that $Z_{kl}^{(0)} = Z_{0,kl}^{(0)}, k,l = 1,2...n$. The subscript kl denotes the kth row and l^{th} column of any matrix appearing in this article. Thus we have:

$$Z^{(0)} = \begin{bmatrix} Z_{0,11}^{(0)} & \cdots & Z_{0,1l}^{(0)} & \cdots & Z_{0,1n}^{(0)} & Z_{1r}^{(0)} \\ \vdots & \ddots & \vdots & \ddots & \vdots & \vdots \\ Z_{0,k1}^{(0)} & \cdots & Z_{0,kl}^{(0)} & \cdots & Z_{0,kn}^{(0)} & Z_{kr}^{(0)} \\ \vdots & \ddots & \vdots & \ddots & \vdots & \vdots \\ Z_{0,n1}^{(0)} & \cdots & Z_{0,nl}^{(0)} & \cdots & Z_{0,nn}^{(0)} & Z_{nr}^{(0)} \\ Z_{r1}^{(0)} & \cdots & Z_{rl}^{(0)} & \cdots & Z_{rn}^{(0)} & Z_{rr}^{(0)} \end{bmatrix} \qquad (1)$$

Therefore, to fully obtain $Z^{(0)}$, only its last row and column need to be calculated. In order to derive $Z_{kr}^{(0)}$ ($k = 1,2...n$), inject a current source of 1 Ampere into a single bus k ($k = 1,2...n$) as shown in Figure 3. i_1, i_2 denote the currents flowing from bus p to q and r respectively in Figure 3.

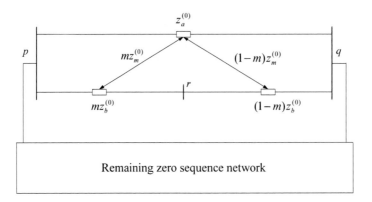

Figure 2. The zero sequence network with an additional fault bus

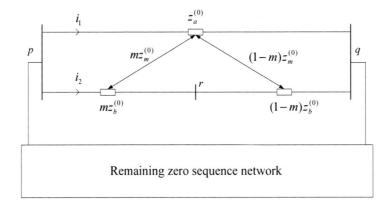

Figure 3. Network with 1 A injected to a single bus k.

Making use of the bus impedance matrix in (1), the voltages at bus p, q and r in Figure 3 will be

$$V_p = Z_{0,pk}^{(0)}$$

$$V_q = Z_{0,qk}^{(0)}$$

$$V_r = Z_{rk}^{(0)} \tag{2}$$

From Figure 3, the following three equations hold:

$$V_p - V_r = mz_b^{(0)}i_2 + mz_m^{(0)}i_1 \tag{3}$$

$$V_r - V_q = (1-m)z_b^{(0)}i_2 + (1-m)z_m^{(0)}i_1 \tag{4}$$

$$V_p - V_q = z_a^{(0)} i_1 + z_m^{(0)} i_2 \tag{5}$$

Solving (3)~(5), the following arrives:

$$V_r = V_p - m(V_p - V_q) \tag{6}$$

Substituting (2) into (6), it is obtained that

$$Z_{rk}^{(0)} = Z_{0,pk}^{(0)} - m(Z_{0,pk}^{(0)} - Z_{0,qk}^{(0)}), \ k = 1,2 \ldots n \tag{7}$$

Define

$$B_k^{(0)} = Z_{0,pk}^{(0)} \tag{8}$$

$$C_k^{(0)} = Z_{0,qk}^{(0)} - Z_{0,pk}^{(0)} \tag{9}$$

Then we have

$$Z_{rk}^{(0)} = B_k^{(0)} + C_k^{(0)} m, \ k = 1,2 \ldots n \tag{10}$$

For the derivation of $Z_{rr}^{(0)}$, inject one current source of 1 Ampere into bus r as shown in Figure 4. i_1, i_2 denote the currents flowing from p to q and r respectively in Figure 4.

Based on the bus impedance matrix in (1), the voltages at bus p, q and r in Figure 4 are

$$V_p = Z_{pr}^{(0)}$$

$$V_q = Z_{qr}^{(0)}$$

$$V_r = Z_{rr}^{(0)} \tag{11}$$

From Figure 4, the following equations hold:

$$V_p - V_r = m z_m^{(0)} i_1 + m z_b^{(0)} i_2 \tag{12}$$

$$V_r - V_q = (1-m) z_b^{(0)} (i_2 + 1) + (1-m) z_m^{(0)} i_1 \tag{13}$$

$$V_p - V_q = z_a^{(0)} i_1 + m z_m^{(0)} i_2 + (1-m) z_m^{(0)} (i_2 + 1) \tag{14}$$

Solving (12)~(14) results in

$$V_r = (1-m)V_p + mV_q + m(1-m)z_b^{(0)}$$

(15)

Substituting (11) into (15), we get

$$Z_{rr}^{(0)} = (1-m)Z_{pr}^{(0)} + mZ_{qr}^{(0)} + m(1-m)z_b^{(0)}$$

(16)

where $Z_{pr}^{(0)}$ and $Z_{qr}^{(0)}$ can be obtained by letting k as p and q in (10) and (16) becomes

$$Z_{rr}^{(0)} = Z_{0,pp}^{(0)} + m(2Z_{0,pq}^{(0)} - 2Z_{0,pp}^{(0)} + z_b^{(0)})$$
$$+ m^2(Z_{0,pp}^{(0)} + Z_{0,qq}^{(0)} - 2Z_{0,pq}^{(0)} - z_b^{(0)})$$

(17)

Define

$$A_0^{(0)} = Z_{0,pp}^{(0)}$$

(18)

$$A_1^{(0)} = 2Z_{0,pq}^{(0)} - 2Z_{0,pp}^{(0)} + z_b^{(0)}$$

(19)

$$A_2^{(0)} = Z_{0,pp}^{(0)} + Z_{0,qq}^{(0)} - 2Z_{0,pq}^{(0)} - z_b^{(0)}$$

(20)

Thus, we attain

$$Z_{rr}^{(0)} = A_0^{(0)} + A_1^{(0)}m + A_2^{(0)}m^2$$

(21)

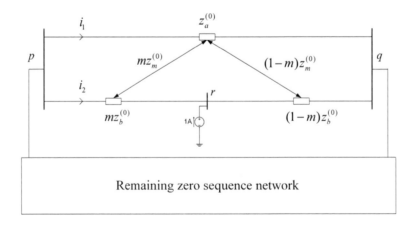

Figure 4. Network with 1 A injected into bus r

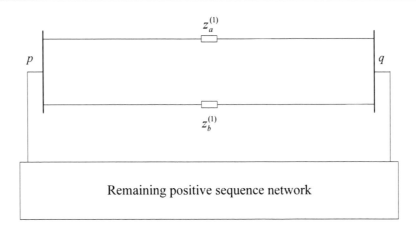

Figure 5. Pre-fault positive sequence network of a sample power system

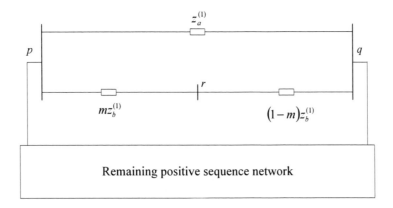

Figure 6. Network with an additional fictitious fault bus for positive sequence network

Let's further consider the construction of bus impedance matrix with addition of a fictitious fault bus for positive and negative sequence networks. The positive sequence network of a sample power system is shown in Figure 5 and the network with an additional fault bus is modeled in Figure 6. $z_a^{(1)}, z_b^{(1)}$ denote the total positive sequence self impedances of the two branches between bus p and q.

The bus impedance matrix of the network in Figure 5 is denoted as $Z_0^{(1)}$ and can be readily developed [10]. $Z^{(1)}$ is the bus impedance matrix for the network in Figure 6 and can be constructed following the same method as for zero sequence network. It is found that the expression of elements of $Z^{(1)}$ share the same form as $Z^{(0)}$. It is assumed that the parameters are the same for positive and negative sequence networks, thus we have $Z^{(2)} = Z^{(1)}$. Here $Z^{(2)}$ represents the bus impedance matrix with an additional fault bus for negative sequence network. Therefore, the bus impedance matrix with addition of a fictitious fault bus for zero-, positive- or negative- sequence network can be written in a compact form as follows:

$$Z_{kl}^{(i)} = Z_{0,kl}^{(i)}, k,l = 1,2...n$$

$$(22)$$

$$Z_{rk}^{(i)} = B_k^{(i)} + C_k^{(i)} m, \ k = 1,2...n \tag{23}$$

$$Z_{rr}^{(i)} = A_0^{(i)} + A_1^{(i)} m + A_2^{(i)} m^2 \tag{24}$$

where $i = 0, 1, 2$ and

$$B_k^{(i)} = Z_{0,pk}^{(i)} \tag{25}$$

$$C_k^{(i)} = Z_{0,qk}^{(i)} - Z_{0,pk}^{(i)} \tag{26}$$

$$A_0^{(i)} = Z_{0,pp}^{(i)} \tag{27}$$

$$A_1^{(i)} = 2Z_{0,pq}^{(i)} - 2Z_{0,pp}^{(i)} + z_b^{(i)} \tag{28}$$

$$A_2^{(i)} = Z_{0,pp}^{(i)} + Z_{0,qq}^{(i)} - 2Z_{0,pq}^{(i)} - z_b^{(i)} \tag{29}$$

This method of obtaining bus impedance matrix with additional fault bus is very clear in concept and computationally efficient. In comparison with the results shown in [9], it can be seen that the formulations of the driving-point impedance of the fault bus and the transfer impedances between this bus and other buses as a function of the fault location take the same form for single-circuit and double-circuit lines. It can be concluded that (22)~(29) are applicable for both single-circuit and double-circuit structures.

II.2. Fault Location Algorithms

At bus k $(k = 1,2...n)$, the following equations hold [10]:

$$E_k^{(1)} = E_k^{(1)0} - Z_{kr}^{(1)} I_f^{(1)} \tag{30}$$

$$E_k^{(2)} = -Z_{kr}^{(2)} I_f^{(2)} \tag{31}$$

$$E_k^{(0)} = -Z_{kr}^{(0)} I_f^{(0)} \tag{32}$$

where

$E_k^{(1)0}$: the pre-fault positive sequence voltage at bus k ;

$E_k^{(0)}, E_k^{(1)}, E_k^{(2)}$: the zero-, positive- and negative-sequence voltage at bus k during the fault respectively;

$I_f^{(0)}, I_f^{(1)}, I_f^{(2)}$: the zero-, positive- and negative-sequence current flowing out of the fault point respectively.

A note of value is that all the sequence voltages and currents are for phase A.

Since the expressions of the driving-point impedance of the fault bus and the transfer impedances between this bus and other buses as a function of the fault location take the same form for single- and double-circuit lines, the overall fault location methods for double-circuit lines are the same as those for single-circuit lines [9].

A. Fault location with measurements from two buses

This section shows the two-bus fault location algorithm based on synchronized voltage measurements from two buses.

Suppose that synchronized pre-fault and fault voltage measurements at bus k and l ($k, l = 1, 2...n$) are available. For bus l, similar to (30), the following formula holds:

$$E_l^{(1)} = E_l^{(1)0} - Z_{lr}^{(1)} I_f^{(1)}$$

(33)

Combining (23), (30) and (33), it follows that

$$\frac{E_k^{(1)} - E_k^{(1)0}}{E_l^{(1)} - E_l^{(1)0}} = \frac{B_k^{(1)} + C_k^{(1)} m}{B_l^{(1)} + C_l^{(1)} m}$$

(34)

Let

$$d_{kl} = \frac{E_k^{(1)} - E_k^{(1)0}}{E_l^{(1)} - E_l^{(1)0}}$$

(35)

The fault location is then derived as

$$m = \frac{B_k^{(1)} - d_{kl} B_l^{(1)}}{d_{kl} C_l^{(1)} - C_k^{(1)}}$$

(36)

Negative sequence or zero sequence voltage measurements, where applicable, can also be employed for fault location. However, positive sequence voltages are preferred due to the fact that no fault type classification is needed.

B. Fault location with measurements from a single bus

This section shows the one-bus fault location algorithm based on voltage measurement from a single bus k ($k = 1, 2...n$). For B to C to ground fault, we have

$$I_f^{(2)} / I_f^{(0)} = (Z_{rr}^{(0)} + 3R_f) / Z_{rr}^{(2)} \tag{37}$$

Using (23), (24), (31) and (32), (37) becomes

$$\frac{E_k^{(2)}}{E_k^{(0)}} = \frac{(B_k^{(2)} + C_k^{(2)}m)(A_0^{(0)} + A_1^{(0)}m + A_2^{(0)}m^2 + 3R_f)}{(B_k^{(0)} + C_k^{(0)}m)(A_0^{(2)} + A_1^{(2)}m + A_2^{(2)}m^2)} \tag{38}$$

The fault location and fault resistance can be obtained by separating (38) into two real equations and solving them.

The fault location methods for other types of faults based on voltage measurement from a single bus are detailed in [9] and are equally applicable to parallel lines and not shown here. Notice fault type classification is necessary before applying the fault location algorithm.

Note that for LLG, LL and LLL faults, multiple solutions are obtained. A solution is defined as a pair of fault location estimate and fault resistance estimate. A solution could be valid solution ($0 \le m \le 1$ and $R_f \ge 0$) or invalid solution ($m < 0$ or $m > 1$ or $R_f < 0$). An invalid solution can be easily identified and removed. In cases where two or more valid solutions arise, one is the true solution and the others are fake solutions. Identification of fake solutions will be illustrated in case studies.

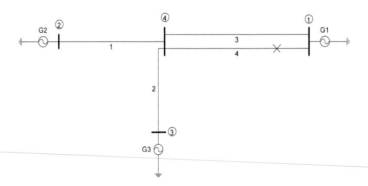

Figure 7. The studied power system diagram

Table I. Transmission Line Data

Line No.	Line Length (km)	Positive/Negative Sequence Impedance (p.u.)	Zero Sequence Impedance (p.u.)
1	178.5	0.015455+j0.116066	0.098871+j0.365188
2	110.2	0.096188+j0.279293	0.243156+j0.822918
3	193.0	0.022172+j0.128174	0.099245+j0.409333
4	193.0	0.022172+j0.128174	0.099245+j0.409333

III. SIMULATION STUDIES

This section presents the simulation results to evaluate the developed fault location algorithms. Electromagnetic Transients Program (EMTP) has been utilized to simulate the studied power system and generate transient waveforms for faults of different types, locations and fault resistances [3]. Discrete Fourier Transform is utilized to extract phasors from the generated waveforms to feed into the developed algorithms to obtain the fault location.

Table II. Generator Data

Gen. No.	Positive/Negative Sequence Impedance (p.u.)	Zero Sequence Impedance (p.u.)
1	0.000600+j0.037343	0.000540+j0.016062
2	0.000900+j0.054236	0.001300+j0.045230
3	0.002200+j0.096514	0.001300+j0.045230

The studied power system is a 230 kV, 100 MVA, 50 Hz transmission line system. The system diagram is shown in Figure 7. The system data are presented in Table I and II. The system is modeled in EMTP based on lumped parameter line model. In this study, shunt capacitance of the line and load are not considered.

The zero-sequence mutual impedance between line 3 and 4 is: $0.079 + j0.2464$ p.u

Table III. Fault Location Result Using Voltages at Two Buses

Fault type	Fault loca. (km)	Fault res. (ohm)	Est. error using volt. at both bus 1 and 2 (%)		Est. error using volt. at both bus 1 and 3 (%)		Est. error using volt. at both bus 1 and 4 (%)	
			Syn.	Unsyn.	Syn.	Unsyn.	Syn.	Unsyn.
AG	30	1	0.040	0.074	0.15	0.20	0.0034	0.0017
		10	0.15	0.19	0.31	0.36	0.0025	0.0050
	90	1	0.051	0.077	0.22	0.26	0.024	0.031
		50	0.48	0.50	0.86	0.89	0.12	0.13
	150	10	0.066	0.078	0.26	0.28	0.094	0.11
		50	0.25	0.26	0.60	0.61	0.28	0.29
BC	30	1	0.0021	0.018	0.046	0.074	0.011	0.010
	90	1	0.0067	0.019	0.066	0.084	0.018	0.021
	150	1	0.0057	0.011	0.053	0.062	0.024	0.030
BCG	30	1	0.0013	0.018	0.038	0.060	0.012	0.011
		50	0.0088	0.029	0.055	0.082	0.014	0.012
	90	10	0.012	0.023	0.064	0.079	0.020	0.023
		50	0.016	0.028	0.074	0.091	0.023	0.025
	150	1	0.0052	0.0095	0.044	0.051	0.025	0.029
		10	0.0077	0.012	0.049	0.056	0.028	0.032
ABC	30	1	0.0015	0.011	0.028	0.041	0.0043	0.0041
	90	1	0.0052	0.011	0.037	0.046	0.0079	0.0098
	150	1	0.0037	0.0062	0.030	0.034	0.011	0.014

From Figure 7, it can be observed that the section between bus 1 and 4 has the double-circuit line structure and the fault occurs on one of the parallel lines, with the cross denoting the fault point. For this particular system, we have $n = 4$, $p = 1$, $q = 4$. The length of the faulted line is 193 km.

The estimation accuracy is evaluated by the percentage error calculated as

$$\%Error = \frac{\left|Actual\ Location - Estimated\ Location\right|}{Total\ Length\ of\ Faulted\ Line} \times 100 \tag{39}$$

where the location of the fault is defined as the distance between the fault point and bus 1.

The developed fault location algorithms are tested under various fault conditions. Table III shows the fault location result produced by two-bus algorithm. The first three columns represent the actual fault type, fault location and fault resistance, respectively. Columns 4-9 indicate the errors of fault location estimate utilizing both synchronized and unsynchronized voltage measurements from two buses.

In Table III, positive sequence voltage measurements are used to carry out two-bus fault location. It can be observed that quite close fault location estimates are produced by using synchronized and unsynchronized data. The fault location results are quite satisfactory. In fact, when synchronized measurements are utilized, the fault location estimate contains an imaginary part, which represents the numerical round off error and is neglected directly. Notice that it is impossible to produce fault location estimate by employing voltage measurements from other bus combinations including 2 and 3, 2 and 4, and 3 and 4 on account of the reason explained in [9].

Table IV presents the one-bus fault location result for AG and BCG faults. Columns 4-7 display the percentage errors of fault location estimate employing the voltage measurements from a single bus. It can be seen that the fault location estimates in Table IV are quite accurate.

The fault location estimate for BCG fault is actually obtained by solving a 4th order polynomial function of m, and the corresponding fault resistance estimate can be solved afterwards. This way, four pairs of fault location and resistance estimates are produced. Any invalid solution if existing can be easily filtered out. It is still possible that two or even more valid solutions remain, where only one solution is true and the rest is fake.

The fake solution can be identified by the following method. We can calculate the voltages of the bus with measurements from all the valid solutions by making use of the bus impedance matrix technique and compare them with the actual voltage measurements. The bus voltages computed from the fake solution differ from the bus voltage measurements, and thus the fake solution can be recognized.

Table V exhibits the one-bus fault location result for BC and ABC faults. Columns 3-14 list the estimated fault location, fault resistance and percentage error in fault location utilizing voltage measurements from a single bus. The actual fault resistance is 1 ohm and the base value of the impedance is 529 ohms.

According to [9], for both BC and ABC faults, a quadratic function with respect to m is formulated to solve for the fault location, which can be used to further calculate the corresponding fault resistance. Two solutions are produced as a result. If one of them is an invalid solution, a unique solution can be obtained. In case two valid solutions are yielded, one of them is a fake solution designated as 'N/A' in Table V. The fake solution identification method proposed for BCG fault fails to distinguish between the two solutions since the computed bus voltages from both of them are the same as the actual bus voltage measurements.

The following approach provides a possibility to identify the fake solution provided that the current measurements at one branch are also available. This way, we can calculate the currents at the branch with measurements from both valid solutions and compare them with the original branch current measurements. If the calculated branch currents are different from the current measurements, the corresponding fault location estimate can be recognized as fake. However, in some cases, both solutions may result in the same currents; in such cases, both fault location estimates will be treated as likely fault locations. Based on our studies, it has been observed when the voltage measurements from any bus of the network and the current measurements from any branch on the faulted double-circuit line are available, the fake fault location estimate can be identified. Specifically, the branch currents may be from the healthy branch of double-circuit line, the branch between bus p and r, or the branch between bus q and r.

For example, in Table V, for a BC fault with fault location of 30 km, based on voltages at bus 2, the algorithm yields two valid solutions: (0.1554, 0.0019) p.u., (0.3878, 0.0028) p.u.. The first element in the bracket represents the fault location estimate and the second one represents the fault resistance estimate. The zero-, positive- and negative-sequence currents on the line from bus 2 to bus 4 calculated from both valid solutions are [0; 0.2331 - 1.6083i; -0.2331 + 1.6083i] p.u., which are the same as the current measurements on this line.

Table IV. Fault Location Result Using Voltages at a Single Bus for AG and BCG Faults

Fault type	Fault loca. (km)	Fault res. (ohm)	Est. err using volt. at bus 1 (%)	Est. err using volt. at bus 2 (%)	Est. err using volt. at bus 3 (%)	Est. err using volt. at bus 4 (%)
AG	30	1	0.016	0.0027	0.0033	0.00034
		10	0.017	0.0026	0.0038	0.0012
	90	1	0.021	0.013	0.014	0.0035
		50	0.022	0.013	0.013	0.0049
	150	10	0.017	0.024	0.027	0.0081
		50	0.018	0.024	0.024	0.0086
BCG	30	1	0.014	0.018	0.017	0.019
		50	0.0028	0.047	0.11	0.031
	90	10	0.027	0.034	0.035	0.016
		50	0.017	0.014	0.048	0.00089
	150	1	0.026	0.14	0.055	0.024
		10	0.025	0.12	0.060	0.014

Table V. Fault Location Result Using Voltages at a Single Bus for BC and ABC Faults

Fault type	Fault loca. (km)	Result using volt. at bus 1			Result using volt. at bus 2			Result using volt. at bus 3			Result using volt. at bus 4		
		Est. fault loca. (p.u.)	Est. fault res. (p.u.)	Est. fault loca. err. (%)	Est. fault loca. (p.u.)	Est. fault res. (p.u.)	Est. fault loca. err. (%)	Est. fault loca. (p.u.)	Est. fault res. (p.u.)	Est. fault loca. err. (%)	Est. fault loca. (p.u.)	Est. fault res. (p.u.)	Est. fault loca. err. (%)
BC	30	0.1556	0.0019	0.016	0.1554	0.0019	0.0040	0.1550	0.0019	0.044	0.1557	0.0019	0.026
					0.3878	0.0028	N/A	0.3883	0.0028	N/A	0.3875	0.0028	N/A
	90	0.4665	0.0020	0.018	0.0994	0.0011	N/A	0.0992	0.0011	N/A	0.0994	0.0011	N/A
					0.4663	0.0020	0.0021	0.4667	0.0020	0.038	0.4663	0.0020	0.0021
	150	0.7778	0.0020	0.060	0.7771	0.0020	0.010	0.7773	0.0020	0.0098	0.7772	0.0020	0.0002
		0.8785	0.0017	N/A									
ABC	30	0.1556	0.0019	0.016	0.1570	0.0018	0.16	0.1584	0.0018	0.30	0.1559	0.0019	0.046
					0.3874	0.0025	N/A	0.3856	0.0025	N/A	0.3890	0.0026	N/A
	90	0.4667	0.0019	0.038	0.1011	0.0011	N/A	0.1018	0.0011	N/A	0.1005	0.0012	N/A
					0.4649	0.0018	0.14	0.4638	0.0017	0.25	0.4661	0.0019	0.022
	150	0.7801	0.0018	0.29	0.7767	0.0018	0.050	0.7763	0.0018	0.090	0.7771	0.0019	0.010
		0.8761	0.0017	N/A									

Thus, the fake fault location estimate can not be identified and there will be two possible fault locations. If the current measurements on the healthy circuit of the faulted double-circuit line are available, one can compute the currents on this branch from both valid solutions. The currents calculated from the first solution are [0; -0.1298 + 0.3036i; 0.1298 - 0.3036i] p.u., which are the same as the current measurements. On the other hand, the same branch currents calculated from the second solution are [0; 0.0012 - 0.4432i; -0.0012 + 0.4432i] p.u., which are different from the current measurements. In this case, the fake solution can be identified and a unique fault location estimate is reached.

In general, the one-bus algorithms are able to yield quite accurate fault location estimate as shown in Table III, IV, and V for different fault types.

IV. CONCLUSION

In this article, fault location algorithms based on voltage measurements applicable to double-circuit lines are developed, which consider the zero-sequence mutual coupling between the parallel lines. The distinctive feature of the method is that only voltage measurements from one or two buses are needed which may be distant from the faulted section. The method for construction of bus impedance matrix with addition of the fault bus is very clear in concept and efficient in computation. Simulation studies have shown that the fault location algorithms can yield quite accurate estimates under various fault conditions. For LL and LLL faults, two possible fault location estimates may be produced. Possible method to identify the fake fault location estimate and its limitations have been discussed.

The developed algorithms are based on lumped parameter line model and the algorithms based on distributed parameter line model are under development and will be reported in the future.

REFERENCES

[1]   Takagi, T; Yamakoshi, Y; Yamaura, M; Kondow, R; Matsushima, T. Development of a new type fault locator using the one-terminal voltage and current data, *IEEE Transactions on Power Apparatus and Systems*, vol. PAS-101 n. 8, August 1982, 2892-2898.

[2]   Eriksson, L; Saha, MM; Rockfeller, GD. An accurate fault locator with compensation for apparent reactance in the fault resistance resulting from remote-end infeed, *IEEE Transactions on Power Apparatus and Systems*, vol. PAS-104 n. 2, February 1985, 424-436.

[3]   Leuven EMTP Centre, Alternative Transient Program, *User Manual and Rule Book*, Leuven, Belgium, 1987.

[4]   Tamer Kawady; Jurgen Stenzel. A practical fault location approach for double circuit transmission lines using single end data, *IEEE Transactions on Power Delivery*, vol. 18 n. 4, October 2003, 1166-1173.

[5]   Liao, Y; Elangovan, S. Digital distance replaying algorithm for first-zone protection for parallel transmission lines, *IEE Proceedings–Part C: Generation, Transmission and*

Distribution, vol. 145 n. 5, September 1998, 531-536.

[6]   Johns, AT; Jamali, S. Accurate fault location technique for power transmission lines, *IEE Proceedings-Generation, Transmission and Distribution*, vol. 137 n. 6, November 1990, 395-402.

[7]   Lawrence, DJ; Cabeza, L; Hochberg, L. Development of an advanced transmission line fault location system part II-Algorithm development and simulation, *IEEE Transactions on Power Delivery*, vol. 7 n. 4, October 1992, 1972-1983.

[8]   Guobing, Song; Jiale, Suonan; Qingqiang, Xu; Ping, Chen; Yozhaong, Ge. Parallel transmission lines fault location algorithm based on differential component net, *IEEE Transactions on Power Delivery*, vol. 20 n. 4, October 2005, 2396-2406.

[9]   Liao, Y. Fault location for single-circuit line based on bus impedance matrix utilizing voltage measurements, *IEEE Transactions on Power Delivery*, vol. 23 n. 2, April 2008, 609-617.

[10] John Grainger and William Stevenson, *Power System Analysis* (New York: McGraw-Hill, Inc., 1994).

INDEX